Ergebnisse der Mathematik und ihrer Grenzgebiete

Band 62

Herausgegeben von P. R. Halmos · P. J. Hilton
R. Remmert · B. Szőkefalvi-Nagy

Unter Mitwirkung von L. V. Ahlfors · R. Baer
F. L. Bauer · A. Dold · J. L. Doob
S. Eilenberg · M. Kneser · G. H. Müller
M. M. Postnikov · B. Segre · E. Sperner

Geschäftsführender Herausgeber: P. J. Hilton

Derek J. S. Robinson

Finiteness Conditions and Generalized Soluble Groups

Part 1

Springer-Verlag Berlin Heidelberg New York
1972

Derek J.S. Robinson
University of Illinois
Urbana, Ill.

AMS Subject Classifications (1970):
Primary 20E15, 20E25, 20E99
Secondary 20F30, 20F35, 20F45, 20F50, 20F99

ISBN 978-3-642-05713-7

To Gabriele

Preface

This book is a study of group theoretical properties of two disparate kinds, firstly finiteness conditions or generalizations of finiteness and secondly generalizations of solubility or nilpotence. It will be particularly interesting to discuss groups which possess properties of both types.

The origins of the subject may be traced back to the nineteen twenties and thirties and are associated with the names of R. Baer, S. N. Černikov, K. A. Hirsch, A. G. Kuroš, O. J. Schmidt and H. Wielandt. Since this early period, the body of theory has expanded at an increasingly rapid rate through the efforts of many group theorists, particularly in Germany, Great Britain and the Soviet Union. Some of the highest points attained can, perhaps, be found in the work of P. Hall and A. I. Mal'cev on infinite soluble groups.

Kuroš's well-known book "The theory of groups" has exercised a strong influence on the development of the theory of infinite groups: this is particularly true of the second edition in its English translation of 1955. To cope with the enormous increase in knowledge since that date, a third volume, containing a survey of the contents of a very large number of papers but without proofs, was added to the book in 1967. Despite this useful addition and the books of M. Hall, E. Schenkman and W. R. Scott, which deal with finite as well as infinite groups, there is a clear need for a detailed account of the theory of finiteness conditions and of generalized soluble and nilpotent groups. The present work represents an attempt to meet this need.

I have sought to collect the most important results in the theory, which are scattered throughout the literature, and to present them in a compact and accessible form with improved and shortened proofs wherever possible. The original aim was to supply full proofs of all the theorems mentioned. However, strict adherence to this rule would probably have doubled the length of the book—it is now possible to write a complete book about locally finite groups for example. Accordingly a compromise has been reached by which certain results are, for one reason or another, stated without proof or, in a few cases, with abbreviated proofs. Of course the choice of what detail to present is in the end a personal one with which the reader may take issue.

Concerning the contents of the book, I might mention that Chapter 1 is of a rather more general nature than is strictly necessary: this has the twofold advantage of supplying a unified basis and of collecting many definitions and elementary results which might otherwise have impeded the development of later chapters. Part of Chapter 2 is also elementary: thereafter the chapters speak for themselves; their admittedly ephemeral aim is to bring the reader abreast of current developments.

Generally speaking, standard notation has been employed. I have followed, albeit reluctantly, the alphabetic notation of Kuroš and Černikov for classes of generalized soluble and nilpotent groups (SN, SI, Z etc.). It would seem a formidable task to find appropriate names for so many classes, although most of these can be rendered symbolically by means of Philip Hall's class operations (which are frequently employed here). However, where suitable terms exist they have been adopted. Thus, following Baer, I prefer "hyperabelian" to "SI^*" and "hypercentral" to "ZA". I might add that the word "series" is used here in contexts where some authors would employ "normal series" or "normal system". Also it seems preferable to speak of a "residual class" rather than a "semi-simple class".

Needless to say, the reader is expected to have a good basic knowledge of the theory of groups, including abelian groups, and of some of the more familiar parts of commutative algebra and ring theory. Otherwise the presentation is, with very few exceptions, self-contained. As a general rule a chapter will depend upon previous chapters in that results established therein may be used.

It is a pleasure to record my gratitude to the many friends and colleagues who have contributed their advice and criticism or who have supplied important information which might otherwise have been passed over. In particular I thank B. Amberg, R. Baer, P. Hall, K. A. Hirsch, K. W. Gruenberg, O. J. Kegel, J. E. Roseblade, S. E. Stonehewer and B. A. F. Wehrfritz. It was at Professor Baer's suggestion that this work was undertaken and I greatly appreciate the immense care and interest which he has taken in reading the manuscript: this has led to numerous improvements. In addition Dr. Roseblade has made many perceptive remarks on the manuscript. My debt to Professor Hall is that of student to teacher; it was through his lectures at Cambridge that I first became interested in the theory of groups and from them that I gained much of my knowledge.

My thanks are due to Springer-Verlag whose patience and cooperation have been all that one would expect of them. In addition I thank the National Science Foundation of the United States and the Department of Mathematics of the University of Illinois in Urbana for support and assistance during the period of writing.

This book is dedicated to my wife Gabriele in recognition of her advice, assistance and often-tried patience, all of which have greatly lightened the labour inherent in such a project as this.

Urbana, Illinois, January 1972 Derek Robinson

Contents

Index of Notation XII

Note to the Reader XV

Chapter 1. Fundamental Concepts in the Theory of Infinite Groups 1

1.1 Group Theoretical Classes and Closure Operations 1
1.2 Series of Subgroups. Classes of Groups Defined by a Series 9
1.3 Radicals and Residuals. Radical and Residual Classes 18
1.4 Finiteness Conditions: A Survey. Finitely Generated and Finitely Presented Groups, Groups with Finite Rank, Periodic and Locally Finite Groups, Maximal and Minimal Conditions 29

Chapter 2. Soluble and Nilpotent Groups 42

2.1 Conjugates and Commutators. Some Generalized Soluble and Nilpotent Groups. Hyperabelian Groups and Hypercentral Groups 42
2.2 Properties of the Upper and Lower Central Series 51
2.3 The Hirsch-Plotkin-Baer Theorem and Related Results. Baer Groups and Gruenberg Groups. Groups with a Category 57

Chapter 3. Maximal and Minimal Conditions 65

3.1 The Maximal and Minimal Conditions on Subgroups. Polycyclic and Černikov Groups. 2-Groups with Max or Min 65
3.2 Groups of Automorphisms of Soluble Groups. Soluble Linear Groups. Theorems of Mal'cev 74
3.3 The Maximal and Minimal Conditions on Abelian Subgroups. Radical Groups with Max-*ab* or Min-*ab* 85
3.4 Locally Finite Groups and Schmidt's Problem. The Hall-Kulatilaka-Kargapolov Theorem. Applications 92

Chapter 4. Finiteness Conditions on Conjugates and Commutators 101

4.1 Central-by-Finite Groups and Applications of Schur's Theorem. Groups with Finite Automorphism Group 101
4.2 Three Problems of Philip Hall on Verbal and Marginal Subgroups. Theorems of Stroud, Turner-Smith and P. Hall 111
4.3 Groups with Finite Conjugacy Classes. Periodic *FC*-Groups. *FC*-nilpotent, *FC*-Hypercentral Groups and *FC*-Soluble Groups 121
4.4 The Classification of Groups with Finite or Černikov Layers: Polovickiĭ's Theorems. An Indecomposable *CL*-Group 133

Chapter 5. Finiteness Conditions on the Subnormal Structure of a Group . . . 143

5.1 Infinite Simple Groups. Embedding Theorems 143
5.2 The Minimal Condition on Normal Subgroups. The Socle. Soluble, Locally Soluble and Locally Nilpotent Groups with Min-*n* 146
5.3 The Maximal Condition on Normal Subgroups. Soluble, Locally Soluble and Locally Nilpotent Groups with Max-*n* 158
5.4 Some Finiteness Conditions on Subnormal Subgroups. Direct Products of Simple Groups, the Minimal Condition on Subnormal Subgroups . 170

Bibliography 187

Author Index 204

Subject Index 207

Contents of Part 2

Chapter 6. Generalized Nilpotent Groups
Chapter 7. Engel Groups
Chapter 8. Local Theorems and Generalized Soluble Groups
Chapter 9. Residually Finite Groups
Chapter 10. Some Topics in the Theory of Infinite Soluble Groups

Index of Notation

Page numbers in brackets refer to the original definition in the text. The symbols I and II refer to Part I and Part II.

(i) Classes and functions

$\mathfrak{X}, \mathfrak{Y}, \mathfrak{Z}, \ldots$	classes of groups (I: 1)
$\mathfrak{X} \leq \mathfrak{Y}$	$\mathfrak{X}$ is a subclass of $\mathfrak{Y}$
$\mathfrak{X}_1 \cdots \mathfrak{X}_n$	extension class (I: 2)
$\mathfrak{X}\mathfrak{Y}_1^{(i_1)} \cdots \mathfrak{Y}_r^{(i_r)}$	(II: 168)
χ	a subgroup theoretical class or property (I: 9)

Special classes of groups

$\mathfrak{F}, \mathfrak{F}_\pi, \mathfrak{P}, \mathfrak{G}, \mathfrak{C}, \mathfrak{A}, \mathfrak{N}, \mathfrak{S}, \mathfrak{T}, \mathfrak{O}$	(I: 1)
$\mathfrak{B}(W)$	(I: 7)
$\mathfrak{B}, \mathfrak{B}_i$	(I: 173)
$\mathfrak{J}$	(I: 174)
$\mathfrak{E}, \mathfrak{E}_n$	(II: 42)
$\mathfrak{U}_s, \mathfrak{U}_{s,n}$	(II: 69)
$\mathfrak{S}_0, \mathfrak{S}_1$	(II: 128, 137)
$\hat{\mathfrak{M}}, \check{\mathfrak{M}}$	(II: 165–166)
Max-f, Min-f	maximal and minimal conditions on f-subgroups (I: 37)
$\boldsymbol{A}, \boldsymbol{B}, \boldsymbol{C}, \ldots$	operations on classes of groups (I: 3)
$\boldsymbol{A} \leq \boldsymbol{B}$	$\boldsymbol{A}\mathfrak{X} \leq \boldsymbol{B}\mathfrak{X}$ for all $\mathfrak{X}$
$\boldsymbol{AB}$	product of operations $\boldsymbol{A}$ and $\boldsymbol{B}$ (I: 3)
$\boldsymbol{A}^\alpha$	ordinal power of $\boldsymbol{A}$ (I: 13)
$\langle \boldsymbol{A}_\lambda : \lambda \in \Lambda \rangle$	closure operation generated by operations $\boldsymbol{A}_\lambda$ (I: 5)

Special operations

$\boldsymbol{I}, \boldsymbol{U}, \boldsymbol{S}, \boldsymbol{S}_n, \boldsymbol{H}, \boldsymbol{P}, \boldsymbol{D}, \boldsymbol{D}_0, \boldsymbol{N}, \boldsymbol{N}_0, \boldsymbol{R}, \boldsymbol{R}_0$	(I: 4)
$\boldsymbol{L}, \boldsymbol{L}_c$	(I: 5)
$\grave{\boldsymbol{P}}, \acute{\boldsymbol{P}}, \hat{\boldsymbol{P}}, \grave{\boldsymbol{P}}_{sn}, \acute{\boldsymbol{P}}_{sn}, \hat{\boldsymbol{P}}_{sn}, \grave{\boldsymbol{P}}_n, \acute{\boldsymbol{P}}_n, \hat{\boldsymbol{P}}_n$	(I: 12–13)
$\acute{\boldsymbol{N}}$	(I: 19)
$\boldsymbol{M}, \overline{\boldsymbol{M}}$	(I: 36, 62)

$\mathfrak{X}^{A}$, $\mathfrak{X}^{-A}$	largest A-closed classes contained in and disjoint from $\mathfrak{X}$ (I: 6–7)
Rad $\mathfrak{X}$, Res $\mathfrak{X}$	the radical and residual classes generated by $\mathfrak{X}$ (I: 20, 23)
Cat $\mathfrak{X}$, $\mathfrak{X}^{(\alpha)}$	classes of groups with an $\mathfrak{X}$-category (I: 36–37)

(ii) Elements and groups

$x, y, z, \ldots$	group elements
1_G, 1	identity element of a multiplicative group
x^y	$y^{-1}xy$
x^{-y}	$(x^y)^{-1}$
x^{y+z}	$x^y x^z$
$[x_1, \ldots, x_n]$	(left normed) commutator (I: 42)
$[x, {}_n y]$	$[x, \underbrace{y, \ldots, y}_{n}]$
$G, H, \ldots$	groups, rings, sets, etc.
$H \leq G$, $H < G$	H is a subgroup and H is a proper subgroup of the group G
$H \lhd G$	H is a normal subgroup of the group G
$H \lhd^{\Sigma} G$	(I: 12)
H *sn* G, H *asc* G, H *desc* G and H *ser* G	H is a subnormal, ascendant, descendant or serial subgroup of G (I 12)
$\langle X_\lambda : \lambda \in \Lambda \rangle$	subgroup generated by the subsets X_λ
$\langle X : R \rangle$	group with generators X and relators R
$X_1 \cdots X_n$, $\prod_{\lambda \in \Lambda} X_\lambda$	products of subsets of a (multiplicative) group
$X_1 \cdots X_n$	additive subgroup generated by all $x_1 \cdots x_n$, $x_i \in X_i$, where X_i is a subset of a ring
X^{-1}	set of all x^{-1}, $x \in X$, where X is a subset of a group
G^n	subgroup generated by all nth powers g^n where $g \in G$ and G is a group
R^n	additive subgroup generated by all $r_1 \cdots r_n$, $r_i \in R$, where R is a ring
$\lvert G \rvert$	cardinality of G
$\lvert G : H \rvert$	index of H in G
$C_G(H)$, $N_G(H)$	centralizer and normalizer of H in G, (where H is a G-operator group)
$\mathrm{Core}_G\, H$	core of H in G (I 60), (where H is a G-operator group)
$\mathrm{Aut}_G\, H$	group of automorphisms of H induced by elements of G (where H is a G-operator group)
Aut G, Inn G	automorphism and inner automorphism group
End G	set of all endomorphisms of G
Hom(G, H)	set of all homomorphisms of G into H
R^*	multiplicative group of units of a ring R
RG	group ring of a group G over a ring R
$H \times K$, $H \oplus K$, $\mathrm{Dr}_{\lambda \in \Lambda} H_\lambda$	direct products or sums
$H \otimes K$	tensor product
$H * K$, $\mathrm{Fr}_{\lambda \in \Lambda} H_\lambda$	free products
$H \wr K$, $\mathrm{Wr}_{\lambda \in \Lambda} H_\lambda$	wreath products (II: 18ff.)
$[X_1, \ldots, X_n]$, $\gamma X Y^n$	commutator subgroup (I: 43) $[X, \underbrace{Y, \ldots, Y}_{n}]$

X^Y	normal closure of X in Y (I: 42)
X^y	$y^{-1}Xy$
$X^{G,\alpha}$	αth term of the series of successive normal closures of X in G (I: 173)
$\{\Lambda_\sigma, V_\sigma : \sigma \in \Sigma\}$	series of order type Σ (I: 9–10)

(iii) Special subgroups

$G' = [G, G]$	derived subgroup of G
$G^{(\alpha)}$	αth term of the derived series of G (I: 45)
$\gamma_\alpha(G)$, $\gamma_\alpha^\chi(G)$	αth terms of the lower central and lower χ-central series (I: 29)
$\zeta(G)$	centre of G
$\zeta_\alpha(G)$, $\zeta_\alpha^\chi(G)$	αth terms of the upper central and upper χ-central series (I: 28)
$\bar{\gamma}(G)$, $\bar{\gamma}^\chi(G)$	hypocentre and χ-hypocentre (I: 29)
$\bar{\zeta}(G)$, $\bar{\zeta}^\chi(G)$	hypercentre and χ-hypercentre (I: 28)
$\varrho_{\mathfrak{X}}(G)$, $\varrho_{\mathfrak{X}}^*(G)$	$\mathfrak{X}$-radical and $\mathfrak{X}$-residual (I: 18)
$W(G)$, $W^*(G)$	verbal and marginal subgroup determined by a word or set of words W (I: 8, 9)
$L(G)$, $\bar{L}(G)$	sets of left and bounded left Engel elements of G (II: 40)
$R(G)$, $\bar{R}(G)$	sets of right and bounded right Engel elements of G (II:40)
$\varrho(G)$, $\bar{\varrho}(G)$	(II: 57)
$\omega(G)$	Wielandt subgroup of G (I: 177)
$\sigma(G)$	upper-finite radical of G (II: 180)

(iv) Miscellaneous

$GF(q)$	Galois field with q elements
$M(n, R)$	ring of all $n \times n$ matrices over the ring R
$GL(n, R)$	general linear group of all invertible $n \times n$ matrices over the ring R
$U(n, R)$	group of all upper unitriangular $n \times n$ matrices over the ring R
$M(\Lambda, F)$	McLain's group (II: 14)
$B(m, n)$	the Burnside group with m generators and exponent n (I: 35)
$\mathfrak{A}(J, \pi)$	a special class of J-modules (II: 146)
$\binom{n}{r}$	binomial coefficient
$[x]$	integral part of x
π'	complementary set of primes to π

Note to the Reader

A decimal classification is used throughout this work. Thus, for example, Theorem 1.29 and Theorem 1.29.1 are respectively the ninth and tenth results in the second section of Chapter 1. The complete bibliography is at the end of Part 2: works referred to in Part 1 appear at the end of Part 1 in the form of an extract from the bibliography; this accounts for gaps in the numbering. References are indicated by numbers in square brackets. As a general rule where the work cited contains a dozen or more pages, a page, section or theorem number is given with the reference. When a result is not ascribed to any author, either it is common knowledge or I am unaware of a source in the literature. Group theoretical classes are denoted by Gothic capitals unless customary usage dictates otherwise. Operations on classes appear as boldface capitals.

Chapter 1

Fundamental Concepts in the Theory of Infinite Groups

1.1 Group Theoretical Classes and Closure Operations

A *group theoretical class* or *class of groups* $\mathfrak{X}$ is a class in the usual sense, consisting of groups, with two additional properties:

(a) $G_1 \simeq G \in \mathfrak{X}$ implies that $G_1 \in \mathfrak{X}$,
(b) $\mathfrak{X}$ contains a trivial (or unit) group.

Since certain natural classes of groups recur frequently, it is convenient to have a short fixed alphabet of classes.

$\mathfrak{F}$ finite groups,	$\mathfrak{C}$ cyclic groups,
$\mathfrak{F}_\pi$ finite π-groups,	$\mathfrak{A}$ abelian groups,
$\mathfrak{P}$ periodic groups,	$\mathfrak{N}$ nilpotent groups,
$\mathfrak{G}$ finitely generated groups,	$\mathfrak{S}$ soluble groups.

The two extreme classes of groups are the *class of all groups*

$$\mathfrak{O}$$

and the *class of trivial groups*

$$\mathfrak{I}.$$

The groups which belong to a class $\mathfrak{X}$ are referred to as *$\mathfrak{X}$-groups*. The group theoretical classes are, of course, partially ordered by inclusion and the notation

$$\mathfrak{X} \leq \mathfrak{Y}$$

will be used to denote the fact that $\mathfrak{X}$ is a group theoretical subclass of the group theoretical class $\mathfrak{Y}$, not merely that $\mathfrak{X}$ is a subclass of $\mathfrak{Y}$.

Sometimes it is preferable to deal with *group theoretical properties* or *properties of groups*: A group theoretical property $\mathscr{P}$ is a property pertaining to groups such that if a group G has $\mathscr{P}$, then every isomorphic image of G has $\mathscr{P}$ and such that at least one unit group has $\mathscr{P}$. The groups which have a given group theoretical property form a group theoretical class and to belong to a given group theoretical class is a group theoretical property. Consequently, there is a one-to-one correspondence between the group theoretical classes and the group theoretical properties; for this reason we will often not distinguish between a group theoretical property and the class of groups that possess it.

The Algebra of Group Theoretical Classes

If $\mathfrak{X}$ and $\mathfrak{Y}$ are two classes of groups, the *extension* or *product class*

$$\mathfrak{X}\mathfrak{Y}$$

is defined as follows: a group G belongs to $\mathfrak{X}\mathfrak{Y}$ if and only if there is a normal subgroup N of G such that $N \in \mathfrak{X}$ and $G/N \in \mathfrak{Y}$. Groups in the class $\mathfrak{X}\mathfrak{Y}$ are called *$\mathfrak{X}$-by-$\mathfrak{Y}$ groups.*

It should be observed that this binary algebraic operation on the class of all classes of groups is neither associative nor commutative. For example, let G be an alternating group of degree 4. Then

$$G \in (\mathfrak{C}\mathfrak{C})\,\mathfrak{C}.$$

However G has no non-trivial normal cyclic subgroups, so

$$G \notin \mathfrak{C}(\mathfrak{C}\mathfrak{C}).$$

On the other hand, the inclusion

$$\mathfrak{X}(\mathfrak{Y}\mathfrak{Z}) \leqq (\mathfrak{X}\mathfrak{Y})\,\mathfrak{Z}$$

is universally valid and, indeed, follows at once from the definition. To avoid excessive use of brackets let us agree to write

$$\mathfrak{X}_1\mathfrak{X}_2\mathfrak{X}_3 = (\mathfrak{X}_1\mathfrak{X}_2)\,\mathfrak{X}_3$$

and in general

$$\mathfrak{X}_1 \cdots \mathfrak{X}_{n-1} = (\mathfrak{X}_1 \cdots \mathfrak{X}_n)\,\mathfrak{X}_{n-1},$$

so that $\mathfrak{X}_1 \cdots \mathfrak{X}_k$ is the most extensive product class that can be formed by bracketing $\mathfrak{X}_1, \ldots, \mathfrak{X}_k$ in the given order. It is clear how to define powers of a class of groups $\mathfrak{X}$:

$$\mathfrak{X}^0 = \mathfrak{I} \quad \text{and} \quad \mathfrak{X}^n = \underbrace{\mathfrak{X} \cdots \mathfrak{X}}_{n} \quad (n > 0).$$

Notice that the associative and commutative laws are not valid even for the powers of a class of groups: the example above shows that $\mathfrak{C}^3 \neq \mathfrak{C}\mathfrak{C}^2$.

Closure Operations

The past three decades have seen the introduction of a very large number of classes of groups and it would be quite impossible to use a systematic alphabet for them. However, one soon observes that many of these classes are obtainable from simpler classes like $\mathfrak{C}$ and $\mathfrak{F}$ by certain uniform procedures. From this observation stems the importance for our purposes of the concept of a *closure operation*, long familiar in topology. The first systematic use of closure operations in group theory occurs in papers of P. Hall [10], [13] although the ideas are implicit in earlier papers of Baer and also in Plotkin [12] (§ 2).

By an *operation* (on the class of all group theoretical classes) we mean a function $\boldsymbol{A}$ assigning to each class of groups $\mathfrak{X}$ a class of groups $\boldsymbol{A}\mathfrak{X}$ subject to the following conditions:

$$\boldsymbol{A}\mathfrak{I} = \mathfrak{I} \tag{1}$$

and

$$\mathfrak{X} \leqq \boldsymbol{A}\mathfrak{X} \leqq \boldsymbol{A}\mathfrak{Y} \tag{2}$$

whenever $\mathfrak{X} \leqq \mathfrak{Y}$. Should it happen that $\mathfrak{X} = \boldsymbol{A}\mathfrak{X}$, the class $\mathfrak{X}$ is said to be $\boldsymbol{A}$-*closed*. By (1) and (2) the classes $\mathfrak{I}$ and $\mathfrak{O}$ are $\boldsymbol{A}$-closed when $\boldsymbol{A}$ is any operation. A partial ordering of operations is defined as follows: $\boldsymbol{A} \leqq \boldsymbol{B}$ means that

$$\boldsymbol{A}\mathfrak{X} \leqq \boldsymbol{B}\mathfrak{X}$$

for every class of groups $\mathfrak{X}$. Products of operations are formed according to the rule

$$(\boldsymbol{AB})\,\mathfrak{X} = \boldsymbol{A}(\boldsymbol{B}\mathfrak{X}).$$

It is evident that, with these definitions, a set of operations that is closed with respect to forming products of its members is a partially ordered semigroup.

An operation $\boldsymbol{A}$ is called a *closure operation* if it is idempotent, that is if

$$\boldsymbol{A} = \boldsymbol{A}^2. \tag{3}$$

If $\boldsymbol{A}$ is a closure operation, then by (2) and (3) the class $\boldsymbol{A}\mathfrak{X}$ is the uniquely determined, smallest $\boldsymbol{A}$-closed class of groups that contains $\mathfrak{X}$. Thus *if* $\boldsymbol{A}$ *and* $\boldsymbol{B}$ *are closure operations,* $\boldsymbol{A} \leqq \boldsymbol{B}$ *if and only if* $\boldsymbol{B}$*-closure invariably implies* $\boldsymbol{A}$*-closure.*

A closure operation can be defined by specifying the classes of groups that are closed. Let $\mathscr{S}$ be a class of classes of groups and suppose that $\mathscr{S}$ contains both $\mathfrak{I}$ and $\mathfrak{O}$ and that every intersection of members of $\mathscr{S}$ belongs to $\mathscr{S}$: for example, $\mathscr{S}$ might consist of the closed classes of a closure operation. $\mathscr{S}$ determines a closure operation $\boldsymbol{A}$ defined as follows: for any class of groups $\mathfrak{X}$, let $\boldsymbol{A}\mathfrak{X}$ be the intersection of all those members of $\mathscr{S}$ that contain $\mathfrak{X}$ (a non-empty class). The $\boldsymbol{A}$-closed classes are precisely the members of $\mathscr{S}$.

Now we will list some of the most commonly met closure operations. In each case the verification of the defining properties-in one or other form-is immediate.

Two closure operations suggest themselves: the *identity closure operation*

$$\boldsymbol{I}$$

defined by $\boldsymbol{I}\mathfrak{X} = \mathfrak{X}$ for all $\mathfrak{X}$, and the *universal closure operation*

$$\boldsymbol{U}$$

defined by $\boldsymbol{U}\mathfrak{X} = \mathfrak{D}$ for all $\mathfrak{X} \neq \mathfrak{I}$ and $\boldsymbol{U}\mathfrak{I} = \mathfrak{I}$.

The Closure Operations S, S_n, H and P

$\mathfrak{X} = \boldsymbol{S}\mathfrak{X}$ if every subgroup of an $\mathfrak{X}$-group is an $\mathfrak{X}$-group.

$\mathfrak{X} = \boldsymbol{S}_n\mathfrak{X}$ if every normal subgroup of an $\mathfrak{X}$-group is an $\mathfrak{X}$-group.

$\mathfrak{X} = \boldsymbol{H}\mathfrak{X}$ if every homomorphic image of an $\mathfrak{X}$-group is an $\mathfrak{X}$-group. ($\boldsymbol{Q}$ is often used instead of $\boldsymbol{H}$.)

$\mathfrak{X} = \boldsymbol{P}\mathfrak{X}$ if every extension of an $\mathfrak{X}$-group by an $\mathfrak{X}$-group is an $\mathfrak{X}$-group, i.e. if $\mathfrak{X} = \mathfrak{X}^2$.

For example $\boldsymbol{S}\mathfrak{X}$ consists of those groups which are isomorphic with a subgroup of an $\mathfrak{X}$-group, that is which are *embeddable* in an $\mathfrak{X}$-group, and $\boldsymbol{P}\mathfrak{X}$ is the class of *poly-$\mathfrak{X}$-groups* or groups which have a series of finite length in which each factor is an $\mathfrak{X}$-group. Clearly

$$\boldsymbol{P}\mathfrak{X} = \bigcup_{i=0,1,2,\ldots} \mathfrak{X}^i.$$

The Closure Operations D, D_0, N, N_0, R and R_0

$\mathfrak{X}$ is *$\boldsymbol{D}$-closed* (*$\boldsymbol{D}_0$-closed*) if the direct product of any collection (any pair) of $\mathfrak{X}$-groups is an $\mathfrak{X}$-group.

$\mathfrak{X}$ is *$\boldsymbol{N}$-closed* *$\boldsymbol{N}_0$-closed*) if the product of any collection (any pair) of normal $\mathfrak{X}$-subgroups is an $\mathfrak{X}$-group.

$\mathfrak{X} = \boldsymbol{R}\mathfrak{X}$ if $N_\lambda \lhd G$ and $G/N_\lambda \in \mathfrak{X}$, $(\lambda \in \Lambda)$, always imply that $G/\bigcap_{\lambda\in\Lambda} N_\lambda \in \mathfrak{X}$: in other words $\mathfrak{X} = \boldsymbol{R}\mathfrak{X}$ means that $\mathfrak{X}$ is closed with respect to forming subcartesian products. The groups in the class $\boldsymbol{R}\mathfrak{X}$ are called *residually $\mathfrak{X}$-groups* and they may be characterized in the following manner: G is a residually $\mathfrak{X}$-group if and only if to each non-trivial element x of G there corresponds a normal subgroup $N(x)$ not containing x such that $G/N(x) \in \mathfrak{X}$.

The finite residual closure operation $\boldsymbol{R}_0$ is defined by the rule $\mathfrak{X} = \boldsymbol{R}_0\mathfrak{X}$ if whenever $G/N_1 \in \mathfrak{X}$ and $G/N_2 \in \mathfrak{X}$, it always follows that $G/N_1 \cap N_2 \in \mathfrak{X}$.

The closure operations S_n and H, N and R and N_0 and R_0 are dual while P is self-dual in the well-known duality between normal subgroup and factor group: this will become more apparent in the context of radical and residual classes (Section 1.3).

The Local Closure Operations L and L_c

If $\mathfrak{X}$ is a class of groups, $L\mathfrak{X}$ is the class of *locally $\mathfrak{X}$-groups*, consisting of all groups G such that every finite subset of G is contained in an $\mathfrak{X}$-subgroup. This use of the term "locally" is due to D. H. McLain [6]: some authors use the term in a narrower sense (see for example Kuroš [9], Vol. 2, § 55). Occasionally the following closure operation is useful. Let $\mathfrak{c}$ be a cardinal number: the class $L_{\mathfrak{c}}\mathfrak{X}$ is defined to consist of all groups G in which every subset of cardinality at most $\mathfrak{c}$ is contained in an $\mathfrak{X}$-subgroup of G. The operations L_2 and $L_{\aleph_0}$ are of most interest.

The Algebra of Operations

The product of two closure operations need not be a closure operation since it may easily fail to be idempotent. This leads us to make the following definition. Let $\{A_\lambda : \lambda \in \Lambda\}$ be a set of operations (not necessarily closure operations). We define

$$C = \langle A_\lambda : \lambda \in \Lambda \rangle,$$

the closure operation generated by the A_λ, as that closure operation whose closed classes are the classes of groups that are A_λ-closed for every $\lambda \in \Lambda$. It is easily verified that C is the uniquely determined least closure operation such that $A_\lambda \leqq C$ for every $\lambda \in \Lambda$.

Of particular interest are $\langle A \rangle$, *the closure operation generated by the operation* A, and also $\langle A, B \rangle$. In the latter case AB and BA may differ from $\langle A, B \rangle$, even although A and B are closure operations.

By this point it should be clear that the algebra of operations is similar to the algebra of subsets of a group and that $\langle A_\lambda : \lambda \in \Lambda \rangle$ corresponds to $\langle H_\lambda : \lambda \in \Lambda \rangle$, the subgroup generated by the subsets H_λ of a group.

Now follows a simple but useful criterion for the product of two closure operations to be a closure operation.

Lemma 1.11. Let A and B be closure operations. Then AB is a closure operation if and only if $BA \leqq AB$; moreover, if this is the case, $AB = \langle A, B \rangle$.

Proof. Let $BA \leqq AB$. Since $(AB)^2 = A(BA)B \leqq A^2B^2 = AB$, the condition is sufficient. On the other hand, if AB is a closure operation, $A \leqq AB$, $B \leqq AB$ and $AB \leqq \langle A, B \rangle$, so $AB = \langle A, B \rangle$ and $BA \leqq AB$. ▯

Next we give a list of some situations to which the criterion may be applied. In each case the verification is straight forward.

Lemma 1.12. The relation $\boldsymbol{BA} \leqq \boldsymbol{AB}$ is valid in the following cases:

$\boldsymbol{A} = \boldsymbol{P}, \boldsymbol{H}, \boldsymbol{R}, \boldsymbol{R}_0, \boldsymbol{L}$ and $\boldsymbol{B} = \boldsymbol{S}, \boldsymbol{S}_n$,

$\boldsymbol{A} = \boldsymbol{P}, \boldsymbol{L}$ and $\boldsymbol{B} = \boldsymbol{H}$,

$\boldsymbol{A} = \boldsymbol{H}$ and $\boldsymbol{B} = \boldsymbol{R}, \boldsymbol{R}_0$.

Two further relations which are sometimes useful are

$$\boldsymbol{N}_0 \leqq \boldsymbol{PH}$$

and

$$\boldsymbol{RL} \leqq \boldsymbol{LRS}.$$

The first expresses the well-known fact that a class of groups which is closed with respect to forming homomorphic images and extensions is closed with respect to forming finite normal products.

Unary Closure Operations

If G is any group, the isomorphic images of G and the unit groups form a class of groups which will be denoted by

$$(G).$$

A closure operation $\boldsymbol{A}$ is said to be *unary* if for every class $\mathfrak{X}$

$$\boldsymbol{A}\mathfrak{X} = \bigcup_{G \in \mathfrak{X}} \boldsymbol{A}(G).$$

For example $\boldsymbol{S}$, $\boldsymbol{S}_n$ and $\boldsymbol{H}$ are unary, but $\boldsymbol{P}$, $\boldsymbol{R}$ and $\boldsymbol{N}$ are not. *If $\boldsymbol{A}$ is a unary closure operation, the union of any class of $\boldsymbol{A}$-closed classes of groups is $\boldsymbol{A}$-closed.* Consequently, to each class $\mathfrak{X}$ there corresponds a uniquely determined largest $\boldsymbol{A}$-closed subclass of $\mathfrak{X}$, which we denote by

$$\mathfrak{X}^{\boldsymbol{A}}.$$

Naturally the statements $\mathfrak{X} = \boldsymbol{A}\mathfrak{X}$ and $\mathfrak{X}^{\boldsymbol{A}} = \mathfrak{X}$ are equivalent. For example

$$\mathfrak{G}^{\boldsymbol{S}}$$

is the class consisting of groups all of whose subgroups are finitely generated: this is well-known to be the class of groups which satisfy the maximal condition on subgroups.

Another $\boldsymbol{A}$-closed class related to $\mathfrak{X}$ which may be formed whenever $\boldsymbol{A}$ is unary is

$$\mathfrak{X}^{-\boldsymbol{A}}$$

which is defined to be the largest A-closed class of groups that is *disjoint* from $\mathfrak{X}$, i.e., intersects $\mathfrak{X}$ in $\mathfrak{I}$. For example $\mathfrak{F}^{-S}$ is the class of torsion-free groups and $\mathfrak{A}^{-H}$ is the class of *perfect groups*, i.e. groups which coincide with their derived groups. Also

$$\mathfrak{X}^{-I}$$

is the class of all trivial groups and non-$\mathfrak{X}$ groups.

Relative Closure Operations

Closure operations may also be defined relative to a universal class of groups $\mathfrak{U}$ other than $\mathfrak{O}$, the class of all groups; $\mathfrak{F}$ and $\mathfrak{A}$ are two natural candidates here. A closure relation on $\mathfrak{U}$ is defined to be a function A assigning to each subclass $\mathfrak{X}$ of $\mathfrak{U}$ a subclass $A\mathfrak{X}$ of $\mathfrak{U}$ and satisfying the usual three conditions (1), (2) and (3).

Varieties of Groups

Let $W = \{\theta_\alpha \colon \alpha \in A\}$ be a set of words in variables $x_1, x_2, \ldots$, that is to say, W is a subset of the free group on the countable infinite set $\{x_1, x_2, \ldots\}$. Let

$$\mathfrak{B}(W)$$

be the class of all groups G such that each θ_α reduces to the identity element when the variables x_i are replaced by arbitrary elements of G. $\mathfrak{B}(W)$ is called the *variety determined by the set of words* W. For example, the variety determined by the single word $[x_1, \ldots, x_{c+1}]$ is the class of groups that are nilpotent of class at most c.

It is immediate that every variety is S, H, R and L-closed, and a remarkable theorem of Birkhoff from the theory of universal algebras asserts that any S, H and R-closed class of groups is a variety ([2], Theorem 8). A stronger form of Birkhoff's theorem will be proved here: H and R-closure alone are sufficient to ensure that a class of groups is a variety (Kogalovskiĭ [1], [2]; Šain [1]).

Theorem 1.13. A class of groups which is H and R-closed (that is to say is closed with respect to forming homomorphic images and subcartesian products) is a variety.

Proof. Let $\mathfrak{X} = H\mathfrak{X} = R\mathfrak{X}$ and let $\mathfrak{B}$ be the variety determined by the set of words (in a given set of variables) which are identically equal to the unit element in every $\mathfrak{X}$-group. Clearly $\mathfrak{X} \leq \mathfrak{B}$, so it is enough to prove that $\mathfrak{B} \leq \mathfrak{X}$.

If θ is a word which does not belong to the set generating $\mathfrak{B}$, there is an $\mathfrak{X}$-group $H(\theta)$ such that θ is not identically equal to 1 in $H(\theta)$.

Let G be a non-trivial group in $\mathfrak{B}$. Choose ϱ to be an infinite ordinal with cardinality not less than that of G and of each $H(\theta)$, and let F be the free group on a set $\{x_\lambda : \alpha < \varrho\}$. By choice of ϱ there is a homomorphism of F onto G, with kernel M say: then $F/M \simeq G$. Let $u \in F \backslash M$ and suppose that $u = \theta(x_{\lambda_1}, \ldots, x_{\lambda_r})$ where $\alpha_i < \varrho$. Since θ is not identically equal to 1 on G, there is an $\mathfrak{X}$-group $H(\theta)$ such that $\theta(y_{\lambda_1}, \ldots, y_{\lambda_r}) \neq 1$ for some $y_{\lambda_i} \in H(\theta)$. By choice of ϱ, the set $\{x_\lambda : \alpha < \varrho, \alpha \neq \alpha_i, i = 1, \ldots, r\}$ has cardinality equal to that of ϱ, so there is a homomorphism of F onto $H(\theta)$ in which $x_{\alpha_i} \to y_{\lambda_i}$ for $i = 1, \ldots, r$: let K_u denote its kernel. Then $u \notin K_u$ and $F/K_u \simeq H(\theta)$. Now define

$$K = \bigcap_{u \in F \backslash M} K_u .$$

Then $K \lhd F$ and $F/K \in \boldsymbol{R}\mathfrak{X} = \mathfrak{X}$. Clearly $K \leqq M$, so $G \simeq F/M \in \boldsymbol{H}\mathfrak{X} = \mathfrak{X}$ and hence $\mathfrak{B} \leqq \mathfrak{X}$. □

For any class of groups $\mathfrak{X}$ let

$$\boldsymbol{V}\mathfrak{X}$$

be the variety determined by the set of words which are identically equal to $\mathfrak{J}$ in every $\mathfrak{X}$-group. $\boldsymbol{V}\mathfrak{X}$ is called *the variety generated by the class of groups* $\mathfrak{X}$ and evidently $\boldsymbol{V}$ is a closure operation. In fact, by Lemma 1.11, Lemma 1.12 and Theorem 1.13

$$\boldsymbol{V} = \boldsymbol{HR}.$$

Since every variety is $\boldsymbol{S}$ and $\boldsymbol{L}$ closed, it follows that

$$\boldsymbol{S} \leqq \boldsymbol{HR} \quad \text{and} \quad \boldsymbol{L} \leqq \boldsymbol{HR}.$$

The theory of varieties is by now very extensive and no attempt will be made to discuss it. The reader is referred to the book by H. Neumann [1].

Verbal and Marginal Subgroups

Let W be a set of words in variables $x_1, x_2, \ldots$ and let G be any group. Define

$$W(G)$$

to be the subgroup generated by all values of words in W on G. That is to say, $W(G)$ is generated by all $\theta(g_1, \ldots, g_n)$ where $g_i \in G$ and θ is a word in n variables in the set W. Then $W(G)$ is the *verbal subgroup of G determined by W* and the mapping assigning to a group G the verbal subgroup $W(G)$ is called a *verbal mapping*. $W(G)$ is a fully invariant subgroup of G, and in [1] (Theorem 9.1) B. H. Neumann has shown that *every fully invariant subgroup of a free group is verbal.* Evidently G belongs to $\mathfrak{B}(W)$ if and only if $W(G) = 1$.

The dual concept, that of a *marginal subgroup*, was introduced by P. Hall in 1940 [2]. Let N be a normal subgroup of a group G: then N is called *W-marginal* if

$$\theta(g_1, \ldots, g_{i-1}, g_i a, g_{i+1}, \ldots, g_n) = \theta(g_1, \ldots, g_{i-1}, g_i, g_{i+1}, \ldots, g_n)$$

where θ is a word in n variables in the set W, $i = 1, \ldots, n$, $g_i \in G$ and $a \in N$. The following is an equivalent definition: N is W-marginal if whenever $g_i \equiv h_i \bmod N$ and θ is a word in n variables in W, it always follows that $\theta(g_1, \ldots, g_n) = \theta(h_1, \ldots, h_n)$. There is a uniquely determined largest W-marginal subgroup in a group G denoted by

$$W^*(G).$$

Clearly $W^*(G)$ is characteristic in G and G belongs to $\mathfrak{B}(W)$ if and only if $W^*(G) = G$. For example, when W consists of the single word $[x_1, \ldots, x_c]$, the marginal subgroup $W^*(G)$ is the cth term of the upper central series of G, while $W(G)$ is the cth term of the lower central series.

Subgroup Theoretical Classes and Properties

So far only properties of groups as a whole have been considered: however we shall frequently have to deal with properties that relate to subgroups of a group.

Let χ be a property pertaining to subgroups. If H has the property χ when regarded as a subgroup of a group G, we shall write

$$H\chi G.$$

χ is called a *subgroup theoretical property* if $1_G\chi G$ is always valid and if $H^\theta\chi G^\theta$ follows from $H\chi G$ whenever θ is an isomorphism of G with some other group.

Similarly, we define a *subgroup theoretical class* as a class of pairs (H, G) where G is a group and H is a subgroup of G, containing $(1_G, G)$ for all G and containing with (H, G) all (H^θ, G^θ) where θ is an isomorphism of G. These concepts are related in the same way as properties of groups and classes of groups, and it is usually unnecessary to distinguish between them. Examples of subgroup theoretical properties are "normal", "central", "verbal" and "marginal". Clearly it is possible to develop a theory of closure operations on subgroup theoretical classes along the same lines as for classes of groups.

1.2 Series of Subgroups. Classes of Groups Defined by a Series

Let H be a subgroup of a group G and let Σ be a linearly ordered set. A *series between H and G* with order type Σ is a set of subgroups of G

$$\mathscr{S} = \{\Lambda_\sigma, V_\sigma : \sigma \in \Sigma\}$$

such that

(i) each Λ_σ and V_σ contains H.

(ii) $G \backslash H = \bigcup_{\sigma \in \Sigma} (\Lambda_\sigma \backslash V_\sigma)$.

(iii) $\Lambda_\tau \leqq V_\sigma$ if $\tau < \sigma$.

(iv) $V_\sigma \lhd \Lambda_\sigma$.

A series between 1 and G is called a *series in* G. The subgroups Λ_σ and V_σ are the *terms* of the series and the groups $\Lambda_\sigma / V_\sigma$ are the *factors* of the series. It follows from (iii) and (iv) that $\Lambda_\tau \leqq \Lambda_\sigma$ and $V_\tau \leqq V_\sigma$ if $\tau \leqq \sigma$. Hence the terms of $\mathscr{S}$ are linearly ordered by inclusion and the ordering of the Λ_σ and V_σ corresponds to the ordering of Σ.

Let $x \in G \backslash H$: then by (ii)' $x \in \Lambda_\sigma \backslash V_\sigma$ for some $\sigma = \sigma(x) \in \Sigma$. If $x \in \Lambda_\tau$, then $\tau \geqq \sigma$ by (iii). Similarly, if $x \notin V_\tau$, then $\tau \leqq \sigma$. Hence $\sigma(x)$ is simultaneously the least element of Σ such that $x \in \Lambda_{\sigma(x)}$ and the greatest element of Σ such that $x \notin V_{\sigma(x)}$. Clearly $\Lambda_{\sigma(x)}$ is the intersection of all the terms of $\mathscr{S}$ that contain x and $V_{\sigma(x)}$ is the union of all the terms of $\mathscr{S}$ that fail to contain x. More generally we have

$$\Lambda_\sigma = \bigcap_{\tau > \sigma} V_\tau \quad \text{and} \quad V_\sigma = \bigcup_{\tau < \sigma} \Lambda_\tau \tag{4}$$

for all $\sigma \in \Sigma$, provided that these are interpreted as $\Lambda_\sigma = G$ when σ is the last element of Σ and $V_\sigma = H$ when σ is the first element of Σ, should either of these exist.

This definition of a series is due to P. Hall [13]. Closely related is Kuroš's concept of a *normal system*, which antedates Hall's by many years. Let H be a subgroup of a group G: a *normal system* between H and G is a chain of subgroups $\mathscr{S}$ such that (i) if $X \in \mathscr{S}$, then $H \leqq X \leqq G$, (ii) $H \in \mathscr{S}$ and $G \in \mathscr{S}$, (iii) $\mathscr{S}$ is *complete*, that is to say it contains all unions and intersections of its members and (iv) if X has an immediate successor Y in the natural ordering of $\mathscr{S}$, then $X \lhd Y$. Thus a normal system is just a complete series or a series containing all intersections of its Λ-subgroups and all unions of its V-subgroups.

Let $\mathscr{S}$ be a series between H and G and let $\mathscr{S}^*$ be the normal system obtained from $\mathscr{S}$ by adding all intersections and unions of terms of $\mathscr{S}$. Then the terms of $\mathscr{S}^*$ are labelled by the elements of Σ^*, the Dedekind completion of Σ.

Ascending and Descending Series

There are two types of series which are of special interest, *ascending series* and *descending series*, when the linearly ordered set Σ or its reverse Σ'

is well-ordered. Since only the order type is important, a well-ordered set may be replaced by a set of ordinals $\{\beta : \beta < \alpha\}$ for a suitable ordinal α.

Suppose that

$$\Sigma = \{\beta : \beta < \alpha\}$$

and let H be a subgroup of a group G. Let $\mathscr{S} = \{\Lambda_\sigma, V_\sigma : \sigma \in \Sigma\}$ be a series between H and G with order type Σ. By the first equation of (4) we have $\Lambda_\sigma = \bigcup_{\tau > \sigma} V_\tau = V_{\sigma+1}$ if $\sigma + 1 < \alpha$ and $\Lambda_\sigma = G$ if $\sigma + 1 = \alpha$. Hence the Λ_σ are superfluous, with the possible exception of $\Lambda_{\alpha-1}$, which, if it exists, equals G. The completion of the series is obtained by adding $V_\alpha = G$. It is usual to work with the *complete ascending series*

$$H = V_0 \lhd V_1 \lhd \cdots V_\sigma \lhd V_{\sigma+1} \cdots V_\alpha = G$$

(Kuroš calls such a series an ascending normal series, but this term will be used in a different sense here). Observe that if λ is a limit ordinal,

$$V_\lambda = \bigcup_{\beta < \lambda} V_\beta .$$

If Σ is an ordered set, let Σ' denote the reverse of Σ, i.e. the same set but with the reverse ordering $<'$ defined by the rule $a <' b$ if and only if $b < a$. Suppose Σ' is well-ordered, say

$$\Sigma' = \{\beta : \beta < \alpha\},$$

and let $\mathscr{S}$ be a series with order type Σ from H to G. It is more convenient to label the terms of $\mathscr{S}$ by elements of Σ'. With this convention the second equation of (4) becomes $V_\sigma = \bigcup_{\tau > \sigma} \Lambda_\tau = \Lambda_{\sigma+1}$ if $\sigma + 1 \neq \alpha$ and $V_\sigma = H$ if $\sigma + 1 = \alpha$. Hence the V_σ are superfluous, with the possible exception of $V_{\alpha-1}$, which, if it exists, equals H. The completion of $\mathscr{S}$ is obtained by adding $\Lambda_\alpha = H$: thus we obtain the *complete descending series*

$$H = \Lambda_\alpha \cdots \Lambda_{\sigma+1} \lhd \Lambda_\sigma \cdots \Lambda_1 \lhd \Lambda_0 = G$$

(or descending normal series in the Kuroš terminology). The equation

$$\Lambda_\lambda = \bigcap_{\alpha < \lambda} \Lambda_\alpha$$

is valid for limit ordinals λ.

In the future all ascending and descending series will be assumed complete in the above sense. The most familiar series are *series of finite length*, that is to say series with finite order type. Such a series is, of course, both ascending and descending.

Serial, Ascendant, Descendant and Subnormal Subgroups

A subgroup H of a group G is termed *serial* in G if there is a series between H and G. Should the series be ascending (descending), H is said to be *ascendant* (*descendant*) in G, and if the series has finite length, H is said to be *subnormal* in G.* The notations

$$H \text{ ser } G,\ H \text{ asc } G,\ H \text{ desc } G,\ H \text{ sn } G,$$

are self-explanatory and will occasionally be used. These relations are generalizations of normality; the most useful are undoubtedly subnormality, introduced in 1939 by Wielandt [2], and ascendance, which was first studied by Plotkin [7] and Gruenberg [3]: descendant subgroups are the subject of a paper by Heineken [1]. Both ascendance and seriality are implicit in a much earlier paper of Baer [7].

If there is a series between H and G with order type Σ, we write

$$H \lhd^{\Sigma} G.$$

The following results are self-evident.

Lemma 1.21. (i) If $H \lhd^{\Sigma} H_1$ and $H_1 \lhd^{\Sigma_1} G$, then $H \lhd^{\Sigma+\Sigma_1} G$. (ii) If $H \lhd^{\Sigma} G$ and if θ is a homomorphism of G, then $H^{\theta} \lhd^{\Sigma} G^{\theta}$ provided that Σ is well-ordered. (iii) If $H \lhd^{\Sigma} K$ and $H_1 \lhd^{\Sigma_1} K_1$, then $H \cap H_1 \lhd^{\Sigma+\Sigma_1} K \cap K_1$; if Σ' and $(\Sigma_1)'$ are well-ordered, then $H \cap H_1 \lhd^{\Sigma_2} K \cap K_1$ where $\Sigma_2' = \max\{\Sigma', \Sigma_1'\}$.

By (ii) ascendance and subnormality are preserved under homomorphisms, but this is not true for either descendance or seriality, as may easily be seen from the example of the infinite dihedral group, which has a descendant subgroup of order 2.

Series with Special Terms or Factors

A series in a group G is said to be a *normal* (*subnormal*) series if each term of the series is normal (subnormal) in G. The meaning of the terms "ascending or descending, normal or subnormal series" is clear.

Very frequently one wishes to consider groups which have a series all of whose factors belong to a given class of groups $\mathfrak{X}$: such a series is called an *$\mathfrak{X}$-series*. Thus for a given class of groups $\mathfrak{X}$ we can form the classes

$$\hat{\boldsymbol{P}}\mathfrak{X},\ \acute{\boldsymbol{P}}\mathfrak{X} \text{ and } \grave{\boldsymbol{P}}\mathfrak{X}$$

of groups which have an $\mathfrak{X}$-series, an ascending $\mathfrak{X}$-series or a descending $\mathfrak{X}$-series: we can also form the classes

$$\hat{\boldsymbol{P}}_{sn}\mathfrak{X},\ \acute{\boldsymbol{P}}_{sn}\mathfrak{X} \text{ and } \grave{\boldsymbol{P}}_{sn}\mathfrak{X}$$

* Some authors use the terms "accessible" and "subinvariant" for subnormal and ascendant respectively.

of groups which have a subnormal $\mathfrak{X}$-series, an ascending subnormal $\mathfrak{X}$-series or a descending subnormal $\mathfrak{X}$-series. Clearly the six functions

$$\hat{P}, \acute{P}, \grave{P}; \hat{P}_{sn}, \acute{P}_{sn}, \grave{P}_{sn} \tag{5}$$

are closure operations. However if—as it is natural to do—we form the classes

$$\hat{P}_n\mathfrak{X}, \acute{P}_n\mathfrak{X} \text{ and } \grave{P}_n\mathfrak{X}$$

of groups which have a normal $\mathfrak{X}$-series, an ascending normal $\mathfrak{X}$-series or a descending normal $\mathfrak{X}$-series, the functions

$$\hat{P}_n, \acute{P}_n, \grave{P}_n \tag{6}$$

are operations but not closure operations since normality is not a transitive relation. Indeed Kovács and Neumann [2], using results of P. Hall on wreath powers, have constructed examples to show that

$$(\hat{P}_n)^\alpha \mathfrak{A} < (\hat{P}_n)^{\alpha+1} \mathfrak{A}$$

for each ordinal α. The notation here is as follows: if A is an operation on classes of groups,

$$A^0 = I, A^{\alpha+1} = AA^\alpha \text{ and } A^\lambda\mathfrak{X} = \bigcup_{\beta<\lambda} A^\beta\mathfrak{X}$$

for all ordinals α and all limit ordinals λ. The situation is somewhat different for $\acute{P}_n$ and $\grave{P}_n$: these functions are discussed in Section 1.3 (Theorems 1.34 and 1.36). The inclusion relations

$$P \leqq \acute{P} \leqq \hat{P} \text{ and } P \leqq \grave{P} \leqq \hat{P}$$

are obviously valid and corresponding statements hold for the normal and subnormal P-functions. Notice that P is the finite analogue of both $\hat{P}$ and $\hat{P}_{sn}$: the finite analogue of $\hat{P}_n$ is the operation

$$P_n$$

defined as follows: $P_n\mathfrak{X}$ consists of all groups which have a normal $\mathfrak{X}$-series of finite length. Application of the nine functions (5) and (6) to the class of abelian groups yields classes of generalized soluble groups—not all of them distinct—which will be discussed in detail in Chapter 8.

Some important classes of groups, particularly generalized nilpotent groups, cannot be defined in terms of $\mathfrak{X}$-series. It is therefore necessary to broaden the concept.

Let χ be a subgroup theoretical property. A *normal* series $\{\Lambda_\sigma, V_\sigma : \sigma \in \}$ in a group G is called a *χ-series* if for each $\sigma \in \Sigma$

Λ_σ/V_σ has the property χ as a subgroup of G/V_σ.

The meaning of the terms *ascending χ-series* and *descending χ-series* is clear. Groups which have an ascending χ-series are called *hyper-χ groups*. For example, when χ is centrality we obtain the class of *hypercentral* (*or ZA-*) *groups* and when χ is the property of being a normal $\mathfrak{X}$-subgroup —where $\mathfrak{X}$ is a class of groups—we obtain the class $\acute{\boldsymbol{P}}_n\mathfrak{X}$ of *hyper-$\mathfrak{X}$ groups*.

Characterizations by Means of Homomorphic Images

The following characterization of hyper-χ groups is of a type frequently employed by Baer.

Lemma 1.22. Let χ be a subgroup theoretical property which is inherited by homomorphic images, that is to say $X^\theta \chi G^\theta$ holds whenever $X\chi G$ holds and θ is a homomorphism of G into some group. Then a group G is a hyper-χ group if and only if each non-trivial homomorphic image H of G contains a non-trivial normal subgroup K such that $K\chi H$.

The proof is easy and we omit it. For example, the lemma applies to hyper-$\mathfrak{X}$ groups where $\mathfrak{X}$ is any $\boldsymbol{H}$-closed class of groups, and to hypercentral groups.

The classes $\acute{\boldsymbol{P}}\mathfrak{X}$ and $\acute{\boldsymbol{P}}_{sn}\mathfrak{X}$ admit of similar if less trivial characterizations.

Lemma 1.23. (Ščukin [8]). Let $\mathfrak{X}$ be a class of groups that is closed with respect to forming homomorphic images. Then a group has an ascending $\mathfrak{X}$-series (an ascending subnormal $\mathfrak{X}$-series) if and only if every non-trivial homomorphic image of G contains a non-trivial ascendant (subnormal) $\mathfrak{X}$-subgroup.

Proof. For example, consider the case of an ascending $\mathfrak{X}$-series: the proof in the other case is similar but simpler. Suppose that $\{G_\beta: \beta \leqq \alpha\}$ is an ascending $\mathfrak{X}$-series in G and let N be a proper normal subgroup of G. Then for some $\beta < \alpha$, we must have $G_{\beta+1} \not\leqq N$ and $G_\beta \leqq N$. Now $G_{\beta+1}N/N$ is a homomorphic image of $G_{\beta+1}/G_\beta$, so $G_{\beta+1}N/N \in \mathfrak{X}$. Clearly $G_{\beta+1}N/N$ is ascendant in G/N.

Conversely, assume that non-trivial homomorphic images of G contain non-trivial ascendant $\mathfrak{X}$-subgroups. Let $G \neq 1$ and let H be a non-trivial ascendant $\mathfrak{X}$-subgroup of G. Then there is an ascending series

$$H = H_0 < H_1 < \cdots H_\lambda = G.$$

Define

$$\bar{H}_\beta = H^{H_\beta}$$

and observe that $\{\bar{H}_\beta\colon 1 \leqq \beta \leqq \alpha\}$ is an ascending series between H and H^G. Also if $x \in H_{\beta+1}$, then $(\bar{H}_\beta)^x \lhd \bar{H}_{\beta+1}$ and

$$(\bar{H}_\beta)^{H_{\beta+1}} = \bar{H}_{\beta+1}.$$

Therefore $\bar{H}_{\beta+1}$ is a normal product of conjugates of $\bar{H}_\beta$ in $H_{\beta+1}$. Hence, if $\bar{H}_\beta$ has an ascending $\mathfrak{X}$-series, a series of the same kind can be obtained in $\bar{H}_{\beta+1}$ by forming successive products of conjugates of $\bar{H}_\beta$ and refining the resulting series—$\boldsymbol{H}$-closure is used at this point. On the other hand, if λ is a limit ordinal and if for each $\beta < \lambda$ the group $\bar{H}_\beta$ has an ascending $\mathfrak{X}$-series, then an ascending $\mathfrak{X}$-series between $\bar{H}_\beta$ and $\bar{H}_{\beta+1}$ can be constructed for each $\beta < \lambda$. Hence $\bar{H}_\lambda$ has an ascending $\mathfrak{X}$-series. Thus by transfinite induction $\bar{H}_\alpha = H^G \in \acute{\boldsymbol{P}}\mathfrak{X}$. Passing to G/H^G, we can repeat the argument: in this manner an ascending $\mathfrak{X}$-series can be constructed in G. ▯

We remark that the conclusions of Lemma 1.23 can be expressed symbolically. For example, the characterization of $\acute{\boldsymbol{P}}_{sn}\mathfrak{X}$ is equivalent to the validity of

$$\acute{\boldsymbol{P}}_{sn}\boldsymbol{H}\mathfrak{X} = ((\boldsymbol{H}\mathfrak{X})^{-\boldsymbol{S}_n})^{-\boldsymbol{H}} \tag{7}$$

for all classes $\mathfrak{X}$. The characterization of $\acute{\boldsymbol{P}}\mathfrak{X}$ can be expressed by a similar equation.

Lemma 1.23 is also valid for groups with an ascending $\mathfrak{X}$-series in which each term stands in the relation $\lhd^\alpha$ to G for some prescribed ordinal α (Robinson [9], Lemma 5).

Characterizations by Means of Normal Subgroups

If χ is a subgroup theoretical property, a group which possesses a descending χ-series is called a *hypo-χ group*. This concept specializes to that of a *hypo-$\mathfrak{X}$ group* where $\mathfrak{X}$ is a class of groups: the class of hypo-$\mathfrak{X}$ groups is, of course, $\grave{\boldsymbol{P}}_n\mathfrak{X}$.

Dual to Lemma 1.22, and equally easy to prove, is

Lemma 1.24. Let χ be a subgroup theoretical property such that $(N \cap X/N \cap Y)\,\chi(G/N \cap Y)$ is valid whenever X, Y and N are normal in G and $(X/Y)\,\chi(G/Y)$. Then G is a hypo-χ group if and only if each non-trivial normal subgroup N of G has a proper subgroup X such that $X \lhd G$ and $(N/X)\,\chi(G/X)$.

This lemma applies to hypo-$\mathfrak{X}$ groups where $\mathfrak{X}$ is an $\boldsymbol{S}_n$-closed class of groups and to hypocentral groups.

It is natural to look for an analogue of Lemma 1.23 for descending series. Here the situation is somewhat different because of the equation

$$\grave{\boldsymbol{P}}_{sn}\boldsymbol{S}_n = \grave{\boldsymbol{P}}\boldsymbol{S}_n,$$

which follows from Lemma 1.37 and is part of the statement of Theorem 1.36. If $\mathfrak{X}$ is an $\boldsymbol{S}_n$-closed class, then $\grave{\boldsymbol{P}}_{sn}\mathfrak{X} = \grave{\boldsymbol{P}}\mathfrak{X}$, so that a group which has a descending $\mathfrak{X}$-series also has a descending subnormal $\mathfrak{X}$-series.

Lemma 1.25. Let $\mathfrak{X}$ be a class of groups that is closed with respect to forming normal subgroups. Then a group G has a descending subnormal $\mathfrak{X}$-series if and only if each non-trivial subnormal subgroup of G has a non-trivial homomorphic image belonging to $\mathfrak{X}$.

The necessity of the condition is easy to deal with. To prove its sufficiency, form a descending series by taking successive $\mathfrak{X}$-residuals*; this must reach 1 ultimately; now apply the relation $\boldsymbol{R} \leqq \grave{\boldsymbol{P}}_n\boldsymbol{S}_n$ (see Lemma 1.37) or argue directly.

Hence we can state

$$\grave{\boldsymbol{P}}_{sn}\boldsymbol{S}_n\mathfrak{X} = ((\boldsymbol{S}_n\mathfrak{X})^{-\boldsymbol{H}})^{-\boldsymbol{S}_n}, \qquad (8)$$

which is valid for all classes $\mathfrak{X}$.

Composition Series

If $\mathscr{S}$ and $\mathscr{S}_1$ are series in a group G, then $\mathscr{S}_1$ is said to be a *refinement* of $\mathscr{S}$ if each term of $\mathscr{S}$ is also a term of $\mathscr{S}_1$. A series which has no refinements other than itself is called a *composition series* and its factors are called *composition factors*. Evidently every composition series is complete and a complete series is a composition series if and only if each factor of the series contains no proper non-trivial serial subgroups: non-trivial groups without proper non-trivial serial subgroups are said to be *absolutely simple*. Application of Zorn's Lemma to the set of all refinements of a given series yields the result that *every series in a group G has a refinement which is a composition series of G*. In particular this applies to the series $1 \lhd G$, so *every group has at least one composition series*. In general two composition series in a group G need not have isomorphic factors; for example if G is an infinite cyclic group, each of the series $G > G^2 > G^4 > \cdots 1$ and $G > G^3 > G^9 > \cdots 1$ is a composition series. So the Jordan-Hölder Theorem is not valid for composition series in general.

* For the definition see p. 18.

An *ascending composition series* is an ascending series with no proper refinements which are also ascending series. Thus an ascending composition series is an ascending series all of whose factors are *strictly simple*, that is to say, they have no proper non-trivial ascendant subgroups. By means of Zassenhaus' Lemma it may be shown that *any two ascending series in a group G have isomorphic ascending refinements* and hence that *any two ascending composition series of a group G have isomorphic factors* (Kuroš [2]; [9], § 56).

A subnormal series without proper subnormal refinements is called a *subnormal composition series*: such a series must have simple factors. The original form of the Jordan-Hölder Theorem applies to groups with a subnormal composition series of finite length, a class of groups which is investigated in the fundamental paper of Wielandt [2]: see also Berman and Lyubimov [1].

Clearly absolutely simple groups are strictly simple and strictly simple groups are simple. Also it is easy to show that *every finitely generated simple group is absolutely simple.* The first example of a simple group that is not absolutely simple is due to Chehata [1], who constructed an algebraically simple, ordered group. The first example of a simple group that is not strictly simple was given by P. Hall [13]; Hall's construction is discussed in detail in section 8.4. It is not known whether Chehata's group is strictly simple and no example of a strictly simple group that is not absolutely simple has yet been given.

In [11] (Theorem 2.5) Plotkin shows that *if a group has an ascending series with simple factors, then any two such series have isomorphic factors.* This is not a special case of the theorem of Kuroš cited above in view of the existence of non-strictly simple groups. Guterman [1] investigates complete series with the following property: if X and Y are terms of the series and $X < Y$, there is a term of the series Z such that $X \leqq Z \lhd Y$ and $Z \neq Y$. For series of this type the composition series are just the series with simple factors.

Chief Series

A *chief* or *principal series* in a group is a normal series (or invariant series in Kuroš's terminology) which has no normal refinements other itself. A normal series in a group G is a chief series if and only if it is complete and each factor has no proper non-trivial G-admissible subgroups. The factors of a chief series of G are called *chief factors* of G and these are precisely the factors of G with the form H/K where $H < K$, $K \lhd G$ and H/K is a *minimal normal* subgroup of G/K. We stress here that "minimal normal" means "minimal normal non-trivial" and that, in the same spirit, "maximal" and "maximal normal" mean "maximal proper" and "maximal normal proper".

Of course a group may not have any minimal normal subgroups, but it always has chief factors since any normal series can be refined to a chief series. Chief factors are clearly characteristically simple and can in general have a complicated structure.

Ascending chief series are defined in a similar manner: *any two such series in a group G have G-isomorphic factors* (Kuroš [9], § 56, Birkhoff [1]). For further discussion and results see Kuroš and Černikov [1], § 3, Kuroš [9], § 56 and Plotkin [12], § 1.

1.3 Radicals and Residuals

The general theory of radicals in rings and algebras was developed in the 1950's by Amitsur [1] and Kuroš [8]. The concept of a radical and the dual concept of a residual have since been recognized as important in the theory of groups and the first systematic account appears in papers by Kuroš [10], [11]. Radicals are also discussed in the survey paper of Plotkin [12], but there a different definition of a radical class is adopted.

Let $\mathfrak{X}$ be a class of groups and let G be any group: the *$\mathfrak{X}$-radical* of G is the product of all the normal $\mathfrak{X}$-subgroups of G and is denoted by

$$\varrho_{\mathfrak{X}}(G).$$

The nilpotent radical or *Fitting subgroup*, $\varrho_{\mathfrak{N}}(G)$, is perhaps the best known example.

The *$\mathfrak{X}$-residual* of G is the intersection of all normal subgroups of G whose factor groups in G are $\mathfrak{X}$-groups; this is denoted by

$$\varrho^*_{\mathfrak{X}}(G).$$

We remark in passing that $\varrho_{\mathfrak{X}}$ and $\varrho^*_{\mathfrak{X}}$ are examples of group theoretical functions: a function α which assigns to each group G a subgroup $\alpha(G)$ such that

$$(\alpha(G))^\theta = \alpha(G^\theta) \tag{9}$$

for each isomorphism θ of G is called a *group theoretical function*. Observe that $\alpha(G)$ is always characteristic in G. A group theoretical function α is a functor on the category of groups if and only if equation (9) is valid for all homomorphisms θ of G—here the image of a homomorphism $G \to H$ under α is understood to be its restriction to $\alpha(G)$ — and it is not hard to see that this occurs if and only if α is a verbal mapping.

There are many other ways of deriving group theoretical functions from classes of groups and vice versa: see for example the two papers of Baer [42] and [43].

Clearly $\varrho_{\mathfrak{X}}(G) \in \mathfrak{X}$ and $G/\varrho^*_{\mathfrak{X}}(G) \in \mathfrak{X}$ are universally valid if and only if $\mathfrak{X} = \boldsymbol{N}\mathfrak{X}$ and $\mathfrak{X} = \boldsymbol{R}\mathfrak{X}$ respectively. In fact $\boldsymbol{N}$-closure implies the following apparently stronger property (see Zassenhaus [3], p. 246).

Lemma 1.31. If $\mathfrak{X} = \boldsymbol{N}\mathfrak{X}$, then in any group G the $\mathfrak{X}$-radical $\varrho_{\mathfrak{X}}(G)$ contains all subnormal $\mathfrak{X}$-subgroups of G. If in addition the union of any well-ordered ascending chain of ascendant $\mathfrak{X}$-subgroups is an $\mathfrak{X}$-group, then $\varrho_{\mathfrak{X}}(G)$ contains all ascendant $\mathfrak{X}$-subgroups of G.

Proof. Let H be an ascendant $\mathfrak{X}$-subgroup of G where $\mathfrak{X}$ satisfies both hypotheses. Let $H = H_0 \lhd H_1 \lhd \cdots H_\alpha = G$ be an ascending series from H to G and set $\bar{H}^\beta = H^{H_\alpha}$.

Then it is clear that the $\bar{H}_\beta$, $1 \leqq \beta \leqq \alpha$, are the terms of an ascending series from H to H^G. Suppose that $H^G \notin \mathfrak{X}$ and let β be the first ordinal such that $\bar{H}_\beta \notin \mathfrak{X}$. Then β cannot be a limit ordinal, for if it were, $\bar{H}_\beta$ would be the union of a well-ordered ascending chain of ascendant $\mathfrak{X}$-subgroups of G. Hence $\bar{H}_\beta$ is a normal product of conjugates of $\bar{H}_{\beta-1}$ and $\bar{H}_\beta \in \boldsymbol{N}\mathfrak{X} = \mathfrak{X}$. Consequently $H^G \in \mathfrak{X}$ and $H \leqq H^G \leqq \varrho_{\mathfrak{X}}(G)$.

If H is subnormal in G and $\mathfrak{X}$ is merely $\boldsymbol{N}$-closed, the same argument shows that $H \leqq \varrho_{\mathfrak{X}}(G)$. ▯

Corollary 1. If $\mathfrak{X}$ is any class of groups, $\boldsymbol{N}\mathfrak{X}$ is the class of all groups that can be generated by their subnormal $\mathfrak{X}$-subgroups.

Proof. Let $\boldsymbol{A}\mathfrak{X}$ be the class of groups that can be generated by their subnormal $\mathfrak{X}$-subgroups. Then clearly $\boldsymbol{N} \leqq \boldsymbol{A}$. Also, $\boldsymbol{N}$-closure implies $\boldsymbol{A}$-closure by the lemma, so $\boldsymbol{A} \leqq \boldsymbol{N}$ and hence $\boldsymbol{A} = \boldsymbol{N}$. ▯

Corollary 2. If $\mathfrak{X} = \boldsymbol{N}\mathfrak{X} = \boldsymbol{S}_n\mathfrak{X}$ and if H *sn* G, then

$$\varrho_{\mathfrak{X}}(H) = \varrho_{\mathfrak{X}}(G) \cap H.$$

Proof. Since $\mathfrak{X} = \boldsymbol{N}\mathfrak{X}$, we see that $\varrho_{\mathfrak{X}}(H)$ and $\varrho_{\mathfrak{X}}(G)$ belong to $\mathfrak{X}$. Also $\varrho_{\mathfrak{X}}(H) \lhd H$, so $\varrho_{\mathfrak{X}}(H)$ *sn* G and hence $\varrho_{\mathfrak{X}}(H) \leqq \varrho_{\mathfrak{X}}(G) \cap H$ by the lemma. On the other hand, $\varrho_{\mathfrak{X}}(G) \cap H$ *sn* $\varrho_{\mathfrak{X}}(G)$, so $\varrho_{\mathfrak{X}}(G) \cap H \in \boldsymbol{S}_n\mathfrak{X} = \mathfrak{X}$: hence $\varrho_{\mathfrak{X}}(G) \cap H \leqq \varrho_{\mathfrak{X}}(H)$ since $\varrho_{\mathfrak{X}}(G) \cap H \lhd H$. ▯

It is easy to write down a corresponding result for ascendant subgroups. At this point it is natural to introduce the closure operation

$$\acute{\boldsymbol{N}},$$

which is defined as follows: $\acute{\boldsymbol{N}}\mathfrak{X}$ consists of all groups which can be generated by their ascendant $\mathfrak{X}$-subgroups. Clearly

$$\boldsymbol{N}_0 \leqq \boldsymbol{N} \leqq \acute{\boldsymbol{N}}$$

and in fact these inclusions are strict. From the second part of Lemma 1.31 we see that $\boldsymbol{N}$-closure and $\boldsymbol{L}$-closure together imply $\acute{\boldsymbol{N}}$-closure, so $\acute{\boldsymbol{N}} \leqq \langle \boldsymbol{L}, \boldsymbol{N} \rangle$. But it is obvious that $\boldsymbol{N} \leqq \langle \boldsymbol{L}, \boldsymbol{N}_0 \rangle$; thus on combining these results we obtain

$$\acute{\boldsymbol{N}} \leqq \langle \boldsymbol{L}, \boldsymbol{N}_0 \rangle . \tag{10}$$

Radical Classes

A group theoretical class $\mathfrak{X}$ is called a *radical class* if

$$\mathfrak{X} = \boldsymbol{N}\mathfrak{X} = \boldsymbol{H}\mathfrak{X}$$

and if for every group G

$$\varrho_{\mathfrak{X}}(G/\varrho_{\mathfrak{X}}(G)) \text{ is trivial.} \tag{11}$$

In other words $\mathfrak{X}$ is radical if normal products and homomorphic images of $\mathfrak{X}$-groups are $\mathfrak{X}$-groups and if $G/\varrho_{\mathfrak{X}}(G)$ has no non-trivial normal $\mathfrak{X}$-subgroups. This definition is due to Kuroš. Obviously the class of unit groups and the class of all groups are radical classes.

Radical classes can be characterized entirely in terms of closure properties.

Theorem 1.32. (Kuroš [11], § 2). A class of groups is radical if and only if it is $\boldsymbol{N}$, $\boldsymbol{H}$ and $\boldsymbol{P}$-closed, that is to say it is closed with respect to forming normal products, homomorphic images and extensions.

Proof. Suppose that $\mathfrak{X}$ is a radical class and assume that $N \lhd G$, $N \in \mathfrak{X}$ and $G/N \in \mathfrak{X}$. Then $N \leqq \varrho_{\mathfrak{X}}(G)$, so $G/\varrho_{\mathfrak{X}}(G) \in \boldsymbol{H}\mathfrak{X} = \mathfrak{X}$. By (11) we obtain $G = \varrho_{\mathfrak{X}}(G) \in \mathfrak{X}$, so $\mathfrak{X}$ is $\boldsymbol{P}$-closed.

Conversely, let $\mathfrak{X}$ be an $\boldsymbol{N}$, $\boldsymbol{H}$ and $\boldsymbol{P}$-closed class and let G be any group. If $G/\varrho_{\mathfrak{X}}(G)$ has a non trivial normal $\mathfrak{X}$-ubgroup, then, by $\boldsymbol{N}$ and $\boldsymbol{P}$-closure, G would have a normal $\mathfrak{X}$-subgroup properly containing $\varrho_{\mathfrak{X}}(G)$. This is impossible, so $\varrho_{\mathfrak{X}}(G/\varrho_{\mathfrak{X}}(G))$ is trivial and $\mathfrak{X}$ is a radical class. □

The Radical Class Generated by a Class of Groups

It is clear that the intersection of any class of radical classes is itself a radical class (by Theorem 1.32 for example). Consequently, to each class of groups $\mathfrak{X}$ there corresponds a unique minimal radical class that contains $\mathfrak{X}$, the *radical class generated by* $\mathfrak{X}$: this will be denoted by

$$\operatorname{Rad} \mathfrak{X}.$$

One description of Rad $\mathfrak{X}$ is already at hand, for by Theorem 1.32

$$\operatorname{Rad} \mathfrak{X} = \langle \boldsymbol{N}, \boldsymbol{H}, \boldsymbol{P} \rangle \mathfrak{X},$$

so Rad is just the closure operation $\langle \boldsymbol{N}, \boldsymbol{H}, \boldsymbol{P}\rangle$. The following more convenient description of Rad $\mathfrak{X}$ is due to Ščukin [6], [8].

Theorem 1.33. If $\mathfrak{X}$ is any class of groups,

$$\text{Rad}\ \mathfrak{X} = (\acute{\boldsymbol{P}}_{sn}\boldsymbol{H})\ \mathfrak{X}.$$

Proof. Notice first that $\boldsymbol{H}\acute{\boldsymbol{P}}_{sn} \leqq \acute{\boldsymbol{P}}_{sn}\boldsymbol{H}$, so $\langle \acute{\boldsymbol{P}}_{sn}, \boldsymbol{H}\rangle = \acute{\boldsymbol{P}}_{sn}\boldsymbol{H}$, by Lemma 1.11. It is therefore sufficient to prove that

$$\langle \acute{\boldsymbol{P}}_{sn}, \boldsymbol{H}\rangle = \langle \boldsymbol{N}, \boldsymbol{H}, \boldsymbol{P}\rangle.$$

A product of normal $\mathfrak{X}$-subgroups has an ascending normal $\boldsymbol{H}\mathfrak{X}$-series for any class of groups $\mathfrak{X}$. It follows that $\boldsymbol{N} \leqq \langle \acute{\boldsymbol{P}}_{sn}, \boldsymbol{H}\rangle$ and hence that $\langle \boldsymbol{N}, \boldsymbol{H}, \boldsymbol{P}\rangle \leqq \langle \acute{\boldsymbol{P}}_{sn}, \boldsymbol{H}\rangle$. To prove the reverse inclusion let $\mathfrak{X}$ be any $\boldsymbol{N}$, $\boldsymbol{H}$, and $\boldsymbol{P}$-closed class of groups and let $G \in \acute{\boldsymbol{P}}_{sn}\mathfrak{X}$: it is sufficient to prove that $G \in \mathfrak{X}$. Let R be the $\mathfrak{X}$-radical of G; then $R \in \boldsymbol{N}\mathfrak{X} = \mathfrak{X}$ and $G/R \in \boldsymbol{H}\acute{\boldsymbol{P}}_{sn}\mathfrak{X} \leqq \acute{\boldsymbol{P}}_{sn}\boldsymbol{H}\mathfrak{X} = \acute{\boldsymbol{P}}_{sn}\mathfrak{X}$. If $R \neq G$, it follows that G/R has a non-trivial subnormal $\mathfrak{X}$-subgroup; by Lemma 1.31 we conclude that the $\mathfrak{X}$-radical of G/R is non-trivial. The $\boldsymbol{P}$-closure of $\mathfrak{X}$ now yields the contradiction $R < \varrho_{\mathfrak{X}}(G)$. □

Corollary

$$\text{Rad}\ \mathfrak{X} = ((\boldsymbol{H}\mathfrak{X})^{-\boldsymbol{S}_n})^{-\boldsymbol{H}}.$$

This follows via equation (7). Yet another characterization of Rad $\mathfrak{X}$ has been given by Ščukin [6], [8]; it may be stated thus.

Theorem 1.34. If $\mathfrak{X}$ is any class of groups,

$$\text{Rad}\ \mathfrak{X} = (\acute{\boldsymbol{P}}_n)^{\omega+1}\ \boldsymbol{H}\mathfrak{X}.$$

Proof. Obviously $(\acute{\boldsymbol{P}}_n)^{\omega+1} \leqq \acute{\boldsymbol{P}}_{sn}$. By Theorem 1.33 it is enough to prove the following relation:

$$\acute{\boldsymbol{P}}_{sn} \leqq (\acute{\boldsymbol{P}}_n)^{\omega+1}\ \boldsymbol{H}.$$

Let $\mathfrak{X}$ be any class of groups, let G be any group and let S be a subnormal $\mathfrak{X}$-subgroup of G such that $S \lhd^r G$. If $r > 0$, it will be shown that

$$S^G \in (\acute{\boldsymbol{P}}_n)^{r-1}\ \boldsymbol{H}\mathfrak{X}.$$

For $r = 1$ this is obvious, so let $r > 1$. Then $S \lhd^{r-1} \bar{S} = S^G$ and by induction on r we obtain $T = S^{\bar{S}} \in (\acute{\boldsymbol{P}}_n)^{r-2}\ \boldsymbol{H}\mathfrak{X} = \mathfrak{Y}$. If $g \in G$, then

$T^g \lhd \bar{S}$ and $\bar{S}$ is the product of all such T^g. Hence $T^G = \bar{S}$ belongs to the class

$$\acute{\boldsymbol{P}}_n \boldsymbol{H}\mathfrak{Y} = \acute{\boldsymbol{P}}_n \boldsymbol{H}(\acute{\boldsymbol{P}}_n)^{r-2}\boldsymbol{H}\mathfrak{X} \leqq (\acute{\boldsymbol{P}}_n)^{r-1}\, \boldsymbol{H}\mathfrak{X}$$

since $\boldsymbol{H}\acute{\boldsymbol{P}}_n \leqq \acute{\boldsymbol{P}}_n \boldsymbol{H}$.

Assume now that $G \in \acute{\boldsymbol{P}}_{sn}\mathfrak{X}$ and let $1 = G_0 \lhd G_1 \lhd \cdots G_\alpha = G$ be an ascending subnormal $\mathfrak{X}$-series. Let $\bar{G}_\beta = G_\beta^G$ and observe that $G_\beta \lhd^{r(\beta)} G$ for some positive integer $r(\beta)$. By the result of the last paragraph

$$\bar{G}_{\beta+1}/\bar{G}_\beta \in (\acute{\boldsymbol{P}}_n)^{r(\beta+1)-1}\, \boldsymbol{H}\mathfrak{X}.$$

Since the $\bar{G}_\beta$ form an ascending normal series in G,

$$G \in \acute{\boldsymbol{P}}_n((\acute{\boldsymbol{P}}_n)^\omega\, \boldsymbol{H}\mathfrak{X}) = (\acute{\boldsymbol{P}}_n)^{\omega+1}\, \boldsymbol{H}\mathfrak{X},$$

as required. $\square$

If α is an ordinal and $\mathfrak{X}$ is a class of groups, then, following Kuroš and Ščukin, we define *groups of ascending degree α over $\mathfrak{X}$*: the groups of degree 1 are just the $\mathfrak{X}$-groups and if $\alpha > 1$, the groups of degree α are the groups which have an ascending normal series in which each factor has degree less than α. The connection with the function $\acute{\boldsymbol{P}}_n$ is apparent: the groups of ascending degree α over $\mathfrak{X}$ form the class

$$(\acute{\boldsymbol{P}}_n)^{\alpha-1}\, \mathfrak{X} \text{ if } \alpha < \omega$$

and

$$(\acute{\boldsymbol{P}}_n)^{\alpha+1}\, \mathfrak{X} \text{ if } \omega \leqq \alpha.$$

Theorem 1.34 shows that *Rad $\mathfrak{X}$ is the class of all groups with some ascending degree over $\boldsymbol{H}\mathfrak{X}$*; moreover *a group with some ascending degree over $\boldsymbol{H}\mathfrak{X}$ has degree ω* (Ščukin [6], [8]). Notice also that $(\acute{\boldsymbol{P}}_n)^{\omega+1}\boldsymbol{H}$ is a closure operation.

For example Rad $\mathfrak{C}$, the radical class generated by the class of cyclic groups, is

$$\acute{\boldsymbol{P}}_{sn}\mathfrak{C} = (\acute{\boldsymbol{P}}_n)^{\omega+1}\mathfrak{C},$$

a class of generalized soluble groups called *subsoluble groups* (Baer [20], p. 421), Phillips and Combrink [1]). Another description of this class is furnished by Lemma 1.23. It is clear that the classes of cyclic, abelian, nilpotent and soluble groups generate the same radical class.

For corresponding results on radicals in associative rings see Sulinski, Anderson and Divinsky [1].

Residual Classes

A class of groups $\mathfrak{X}$ is called a *residual class* if

$$\mathfrak{X} = \boldsymbol{R}\mathfrak{X} = \boldsymbol{S}_n\mathfrak{X}$$

and if for all groups G

$$\varrho^*_{\mathfrak{X}}(\varrho^*_{\mathfrak{X}}(G)) = \varrho^*_{\mathfrak{X}}(G), \tag{12}$$

that is to say $\varrho^*_{\mathfrak{X}}$ is idempotent. (Kuroš and some other authors us the term *semi-simple class*). The class of unit groups and the class of all groups are residual classes. The simplest characterization of residual classes in terms of closure properties is the following.

Theorem 1.35. (Kuroš [11], § 5). A class of groups is a residual class if and only if it is $\boldsymbol{R}$, $\boldsymbol{S}_n$ and $\boldsymbol{P}$-closed, that is to say it is closed with respect to forming subcartesian products, normal subgroups and extensions.

Proof. If $\mathfrak{X} = \boldsymbol{R}\mathfrak{X} = \boldsymbol{S}_n\mathfrak{X} = \boldsymbol{P}\mathfrak{X}$ and G is any group, then $\varrho^*_{\mathfrak{X}}(G)$ has no non-trivial $\mathfrak{X}$-factor groups: hence $\varrho^*_{\mathfrak{X}}$ is idempotent and $\mathfrak{X}$ is residual. Conversely, let $\mathfrak{X}$ be a residual class and let N and G/N be $\mathfrak{X}$-groups. Then $\varrho^*_{\mathfrak{X}}(G) \leqq N$, so $\varrho^*_{\mathfrak{X}}(G) \in \boldsymbol{S}_n\mathfrak{X} = \mathfrak{X}$: hence $\varrho^*_{\mathfrak{X}}(G) = \varrho^*_{\mathfrak{X}}(\varrho^*_{\mathfrak{X}}(G)) = 1$ and $G \in \mathfrak{X}$. □

The Residual Class Generated by a Class of Groups

The intersection of a class of residual classes is a residual class, so any given class of groups $\mathfrak{X}$ lies in a uniquely determined smallest residual class called the *residual class generated by* $\mathfrak{X}$ and denoted by

$$\text{Res}\ \mathfrak{X}.$$

Theorem 1.35 shows that

$$\text{Res}\ \mathfrak{X} = \langle \boldsymbol{R}, \boldsymbol{S}_n, \boldsymbol{P}\rangle\ \mathfrak{X},$$

so Res is just the closure operation $\langle \boldsymbol{R}, \boldsymbol{S}_n, \boldsymbol{P}\rangle$. To obtain a further description of Res $\mathfrak{X}$ we utilize the function $\grave{\boldsymbol{P}}_n$.

Theorem 1.36. (Čan Van Hao [1]). For any class of groups $\mathfrak{X}$

$$\text{Res}\ \mathfrak{X} = (\grave{\boldsymbol{P}}_n)^2\boldsymbol{S}_n\mathfrak{X} = \grave{\boldsymbol{P}}_{sn}\boldsymbol{S}_n\mathfrak{X} = \grave{\boldsymbol{P}}\boldsymbol{S}_n\mathfrak{X}.$$

The proof depends on the following lemma.

Lemma 1.37. The relations

$$\grave{\boldsymbol{P}} \leqq \grave{\boldsymbol{P}}_n\boldsymbol{R}\boldsymbol{S}_n, \quad \boldsymbol{R}\grave{\boldsymbol{P}}_n \leqq \grave{\boldsymbol{P}}_n\boldsymbol{S}_n$$

and

$$\boldsymbol{R}\hat{\boldsymbol{P}}_n \leqq \hat{\boldsymbol{P}}_n \boldsymbol{S}_n$$

are valid.

Proof. Let $G \in \grave{\boldsymbol{P}}\mathfrak{X}$ and let $1 = G_\lambda \cdots G_2 \lhd G_1 \lhd G_0 = G$ be a descending $\mathfrak{X}$-series in G. Define $R_0 = G$, $R_{\beta+1} = \varrho^*_{\boldsymbol{S}_n\mathfrak{X}}(R_\beta)$ and $R_\lambda = \bigcap_{\gamma<\lambda} R_\gamma$ for all ordinals β and all limit ordinals λ. Then $R_\beta/R_{\beta+1} \in \boldsymbol{R}\boldsymbol{S}_n\mathfrak{X}$. It is sufficient to prove that $R_\beta \leqq G_\beta$, for this implies that $R_\lambda = 1$ and $G \in \grave{\boldsymbol{P}}_n\boldsymbol{R}\boldsymbol{S}_n\mathfrak{X}$. Let β be the first ordinal such that $R_\beta \nleqq G_\beta$. Then β is not a limit ordinal and $R_{\beta-1}G_\beta/G_\beta \lhd G_{\beta-1}/G_\beta \in \mathfrak{X}$, so $R_{\beta-1}/R_{\beta-1} \cap G_\beta \in \boldsymbol{S}_n\mathfrak{X}$. Therefore $R_\beta = \varrho^*_{\boldsymbol{S}_n\mathfrak{X}}(R_{\beta-1}) \leqq R_{\beta-1} \cap G_\beta$, contrary to assumption.

To prove the second relation let G be a non-unit group in the class $\boldsymbol{R}\grave{\boldsymbol{P}}_n\mathfrak{X}$ and let the non-trivial elements of G be well-ordered in some manner, say as $\{x_\beta : \beta < \alpha\}$ where α is an ordinal. For each $\beta < \alpha$ there is a normal subgroup N_β of G such that $x_\beta \notin N_\beta$ and $G/N_\beta \in \grave{\boldsymbol{P}}_n\mathfrak{X}$. Now let $G_0 = G$ and define G_β to be the intersection of all the G_γ for $\gamma < \beta$. Then $1 = G_\alpha \cdots G_2 \leqq G_1 \leqq G_0 = G$ is a descending normal series in G. Since $G_\beta/G_{\beta+1}$ and $G_\beta N_\beta/N_\beta$ are isomorphic as G-operator groups, there exists a descending G-admissible series between $G_{\beta+1}$ and G_β with factors in $\boldsymbol{S}_n\mathfrak{X}$. Hence $G \in \grave{\boldsymbol{P}}_n\boldsymbol{S}_n\mathfrak{X}$. The same argument deals with the third relation. ▯

Corollary. If $\mathfrak{X} = \boldsymbol{S}_n\mathfrak{X}$, then $\grave{\boldsymbol{P}}_n\mathfrak{X}$ and $\hat{\boldsymbol{P}}_n\mathfrak{X}$ are $\boldsymbol{R}$-closed.

Proof of Theorem 1.36. By Lemma 1.37 we have $\boldsymbol{R} \leqq \grave{\boldsymbol{P}}_n\boldsymbol{S}_n$, which implies that $\langle \boldsymbol{R}, \boldsymbol{S}_n, \boldsymbol{P}\rangle \leqq \langle \grave{\boldsymbol{P}}, \boldsymbol{S}_n\rangle = \grave{\boldsymbol{P}}\boldsymbol{S}_n$. Also $\grave{\boldsymbol{P}} \leqq \grave{\boldsymbol{P}}_n\boldsymbol{R}\boldsymbol{S}_n$, by the same lemma, so $\grave{\boldsymbol{P}}\boldsymbol{S}_n \leqq \grave{\boldsymbol{P}}_n\boldsymbol{R}\boldsymbol{S}_n$ and, using $\boldsymbol{R} \leqq \grave{\boldsymbol{P}}_n\boldsymbol{S}_n$ again, we obtain $\grave{\boldsymbol{P}}\boldsymbol{S}_n \leqq (\grave{\boldsymbol{P}}_n)^2\boldsymbol{S}_n$. Hence

$$\langle \boldsymbol{R}, \boldsymbol{S}_n, \boldsymbol{P}\rangle \leqq (\grave{\boldsymbol{P}}_n)^2\boldsymbol{S}_n \leqq \grave{\boldsymbol{P}}_{sn}\boldsymbol{S}_n \leqq \grave{\boldsymbol{P}}\boldsymbol{S}_n.$$

In view of Theorem 1.35 we need only prove the reverse relation $\grave{\boldsymbol{P}}\boldsymbol{S}_n \leqq \langle \boldsymbol{R}, \boldsymbol{S}_n, \boldsymbol{P}\rangle$. Let $\mathfrak{X}$ be an $\boldsymbol{R}$, $\boldsymbol{S}_n$ and $\boldsymbol{P}$-closed class and let $G \in \grave{\boldsymbol{P}}\mathfrak{X}$. Denote the $\mathfrak{X}$-residual of G by R; then $G/R \in \mathfrak{X}$ and

$$R \in \boldsymbol{S}_n\grave{\boldsymbol{P}}\mathfrak{X} \leqq \grave{\boldsymbol{P}}\boldsymbol{S}_n\mathfrak{X} = \grave{\boldsymbol{P}}\mathfrak{X}.$$

If $R \neq 1$, it follows that $S = \varrho^*_{\mathfrak{X}}(R) < R$. But $S \lhd G$ and $G/S \in \boldsymbol{P}\mathfrak{X} = \mathfrak{X}$, so $R = \varrho^*_{\mathfrak{X}}(G) \leqq S$ and $R = S$. By this contradiction $R = 1$ and $G \in \mathfrak{X}$, as required. ▯

Corollary.

$$\operatorname{Res} \mathfrak{X} = ((\boldsymbol{S}_n \mathfrak{X})^{-\boldsymbol{H}})^{-\boldsymbol{S}_n}$$

This follows via equation (8).

One consequence of Theorem 1.36 is that $(\grave{\boldsymbol{P}}_n)^2 \boldsymbol{S}_n$ is a closure operation. If the class of groups of *descending degree* α *over* $\mathfrak{X}$ is defined in the natural way (cf. p. 22), then Res $\mathfrak{X}$ coincides with the class of groups of some descending degree over $\boldsymbol{S}_n \mathfrak{X}$ and all such groups have degree 3 (Čan Van Hao [1]).

It is easy to see that

$$\operatorname{Res} \mathfrak{S} = \operatorname{Res} \mathfrak{N} = \operatorname{Res} \mathfrak{A} = \grave{\boldsymbol{P}} \mathfrak{A} = \grave{\boldsymbol{P}}_n \mathfrak{A},$$

the class of *hypoabelian groups*. However Res $\mathfrak{C}$ is a narrower class since it contains no quasicyclic groups.

Two further remarks about residual classes are in order. If $\mathfrak{X}$ is any $\boldsymbol{S}_n$-closed class, then

$$\hat{\boldsymbol{P}} \mathfrak{X}$$

is a residual class since it is $\grave{\boldsymbol{P}}$ and $\boldsymbol{S}_n$-closed. Thus, for example, $\hat{\boldsymbol{P}} \mathfrak{A}$, the class of *SN-groups*, is a residual class (Livčak [2]). Secondly, *every residual class is closed with respect to forming free products* (Šmel'kin [2]). This may be deduced from Theorem 1.36 and the theorem of Hall and Hartley ([1], Theorem D3) that a free product has a descending series each of whose factors is isomorphic with one of the free factors.

The Kuroš Correspondence between Radical and Residual Classes

Kuroš [11] has established a one-to-one correspondence between radical and residual classes. We shall deduce this from

Theorem 1.38. (i) A class of groups $\mathfrak{X}$ is a radical class if and only if $\mathfrak{X} = \mathfrak{Y}^{-\boldsymbol{H}}$ for some $\boldsymbol{S}_n$-closed class $\mathfrak{Y}$.

(ii) A class of groups $\mathfrak{Y}$ is a residual class if and only if $\mathfrak{Y} = \mathfrak{X}^{-\boldsymbol{S}_n}$ for some $\boldsymbol{H}$-closed class $\mathfrak{X}$.

For the proof we shall require the following lemma

Lemma 1.39. If $\boldsymbol{A}$ and $\boldsymbol{B}$ are unary closure operations and $\mathfrak{X}$ is an $\boldsymbol{A}$-closed class of groups, then

$$((\mathfrak{X}^{-\boldsymbol{B}})^{-\boldsymbol{A}})^{-\boldsymbol{B}} = \mathfrak{X}^{-\boldsymbol{B}}.$$

Proof. Since $(\mathfrak{X}^{-\boldsymbol{B}})^{-\boldsymbol{A}} \leqq (\mathfrak{X}^{-\boldsymbol{B}})^{-\boldsymbol{I}}$, we have

$$((\mathfrak{X}^{-\boldsymbol{B}})^{-\boldsymbol{A}})^{-\boldsymbol{B}} \geqq ((\mathfrak{X}^{-\boldsymbol{B}})^{-\boldsymbol{I}})^{-\boldsymbol{B}} = (\mathfrak{X}^{-\boldsymbol{B}})^{\boldsymbol{B}} = \mathfrak{X}^{-\boldsymbol{B}}.$$

Also $\mathfrak{X}^{-B} \leqq \mathfrak{X}^{-I}$, so $(\mathfrak{X}^{-B})^{-A} \geqq (\mathfrak{X}^{-I})^{-A} = \mathfrak{X}^{A} = \mathfrak{X}$ and therefore

$$((\mathfrak{X}^{-B})^{-A})^{-B} \leqq \mathfrak{X}^{-B},$$

which completes the proof. □

Proof of Theorem 1.38. If $\mathfrak{X}$ is a radical class, then by the Corollary to Theorem 1.33

$$\mathfrak{X} = \operatorname{Rad} \mathfrak{X} = ((\boldsymbol{H}\mathfrak{X})^{-\boldsymbol{S}_n})^{-\boldsymbol{H}} = (\mathfrak{X}^{-\boldsymbol{S}_n})^{-\boldsymbol{H}}, \tag{13}$$

and we can take $\mathfrak{Y} = \mathfrak{X}^{-\boldsymbol{S}_n}$. Conversely, suppose that $\mathfrak{X} = \mathfrak{Y}^{-\boldsymbol{H}}$ for some $\boldsymbol{S}_n$-closed class $\mathfrak{Y}$. Then

$$\operatorname{Rad} \mathfrak{X} = ((\boldsymbol{H}\mathfrak{X})^{-\boldsymbol{S}_n})^{-\boldsymbol{H}} = ((\mathfrak{Y}^{-\boldsymbol{H}})^{-\boldsymbol{S}_n})^{-\boldsymbol{H}} = \mathfrak{Y}^{-\boldsymbol{H}} = \mathfrak{X},$$

by Lemma 1.39. Hence $\mathfrak{X}$ is a radical class, which proves (i). The proof of (ii) is similar. □

Theorem 1.39.1. If $\mathfrak{X}$ is a radical class, $\mathfrak{X}^{-\boldsymbol{S}_n}$ is a residual class, and if $\mathfrak{Y}$ is a residual class, $\mathfrak{Y}^{-\boldsymbol{H}}$ is a radical class. Moreover, the mappings $\mathfrak{X} \to \mathfrak{X}^{-\boldsymbol{S}_n}$ and $\mathfrak{Y} \to \mathfrak{Y}^{-\boldsymbol{H}}$ are inverses of each other and determine a one-one correspondence between the radical classes and the residual classes.

Proof. The first part follows at once from Theorem 1.38. The mappings are inverses by (13) and the corresponding equation $\mathfrak{Y} = (\mathfrak{Y}^{-\boldsymbol{H}})^{-\boldsymbol{S}_n}$. □

Finally, we note an alternative formulation of the Kuroš correspondence due to P. Hall.

Theorem 1.39.2. The following statements about the group theoretical classes $\mathfrak{X}$ and $\mathfrak{Y}$ are equivalent.

(i) $\mathfrak{X} = \boldsymbol{N}\mathfrak{X}$, $\mathfrak{Y} = \boldsymbol{R}\mathfrak{Y}$ and $\varrho_{\mathfrak{X}} = \varrho^{*}_{\mathfrak{Y}}$.

(ii) $\mathfrak{X}$ is a radical class and $\mathfrak{Y} = \mathfrak{X}^{-\boldsymbol{S}_n}$.

(iii) $\mathfrak{Y}$ is a residual class and $\mathfrak{X} = \mathfrak{Y}^{-\boldsymbol{H}}$.

We leave the easy verification to the reader.

If $\mathfrak{X}$ and $\mathfrak{Y}$ are arbitrary classes, $\mathfrak{X}^{-\boldsymbol{S}_n}$ is called the class of $\mathfrak{X}$-*semisimple groups* and $\mathfrak{Y}^{-\boldsymbol{H}}$ the class of $\mathfrak{Y}$-*perfect groups*. When $\mathfrak{X} = \mathfrak{A}$, the terms *semisimple** (having no non-trivial subnormal abelian subgroups) and *perfect* (coinciding with the derived subgroup) are used. Thus in the correspondence between radical and residual classes subsoluble groups correspond to semisimple groups and perfect groups correspond to hypoabelian groups.

It is of some interest that $\mathfrak{I}$ and $\mathfrak{O}$ *are the only classes that are both radical and residual* and hence the only fixed points in the Kuroš corres-

* Sometimes semisimple is taken to mean the absence of non-trivial normal abelian subgroups. The two usages are identical for finite groups.

pondence. For let $\mathfrak{X} \neq \mathfrak{I}$ be both radical and residual. Then $\mathfrak{X}$ is a variety by Theorem 1.13, and hence is $\boldsymbol{S}$-closed. By $\boldsymbol{S}$, $\boldsymbol{H}$ and $\boldsymbol{P}$-closure we conclude that $\mathfrak{X}$ contains all finite p-groups for some prime p. But all free groups are residually finite-p for each prime p (Theorem 9.11). Thus $\mathfrak{X}$ contains all free groups and therefore all groups by $\boldsymbol{H}$-closure.

Radicals and Residuals in Particular Classes of Groups

It is possible to extend the foregoing theory of radicals and residuals by taking a class of groups $\mathfrak{U}$ as the universal class instead of $\mathfrak{D}$: all closure operations must now be taken relative to $\mathfrak{U}$. The classes $\mathfrak{F}$ and $\mathfrak{A}$ are natural candidates and the problem of determining all radical, and therefore all residual, classes on these classes and some of their subclasses has been studied by several authors. For example, let $\mathfrak{X}$ be a radical class on the class of abelian p-groups. If $\mathfrak{X}$ contains a group of order p, then by relative $\boldsymbol{P}$-closure $\mathfrak{X}$ contains all cyclic p-groups and by relative $\boldsymbol{N}$-closure, $\mathfrak{X}$ is the class of all abelian p-groups. If $\mathfrak{X} \neq \mathfrak{I}$ and $\mathfrak{X}$ contains no group of order p, then by $\boldsymbol{H}$-closure $\mathfrak{X}$ contains a group of type p^∞. Hence $\mathfrak{X}$ is just the class of radicable abelian p-groups: the corresponding residual class consists of all reduced abelian p-groups. Hence there are exactly three radical classes and three residual classes on the class of abelian p-groups. In the same way one easily shows that the radical classes on the class of abelian π-groups correspond to all ordered partitions (π_1, π_2, π_3) of the set of primes π into three subsets, some of which may be empty: the radical class corresponding to (π_1, π_2, π_3) consists of all direct products of π_1-groups and radicable π_2-groups. These and further results on radicals in classes of abelian groups may be found in papers of Kuroš [11] and S. E. Dickson [1]: see also Balcerzyk [1], [2].

It is possible for a class other than $\mathfrak{I}$ and $\mathfrak{F}$ to be both a radical and residual class on $\mathfrak{F}$: for example the class of finite soluble groups. Let $\mathfrak{X}$ be a class of finite groups which is both radical and residual on $\mathfrak{F}$ and let $\mathfrak{X}_0$ be the class of all unit groups and all composition factors of $\mathfrak{X}$-groups. Since $\mathfrak{X} = \boldsymbol{H}\mathfrak{X} = \boldsymbol{S}_n\mathfrak{X}$, we have $\mathfrak{X}_0 \leqq \mathfrak{X}$, and by $\boldsymbol{P}$-closure $\boldsymbol{P}\mathfrak{X}_0 \leqq \mathfrak{X}$. But clearly $\mathfrak{X} \leqq \boldsymbol{P}\mathfrak{X}_0$, so in fact $\mathfrak{X} = \boldsymbol{P}\mathfrak{X}_0$. Consequently there is a one-to-one correspondence between the classes of finite groups that are both radical and residual on $\mathfrak{F}$ and the classes of finite simple groups (Kuroš [11]). This is in contrast to the situation when the universal class is the class of all groups.

The χ-Centre and χ-Hypercentre

If χ is a subgroup theoretical property and G is any group, the *χ-centre* of G

$$\zeta^\chi(G)$$

is the subgroup generated by all subgroups of G which have the property χ. If $\mathfrak{X}$ is a class of groups and χ is the property of being a normal $\mathfrak{X}$-subgroup, then $\zeta^\chi = \varrho_{\mathfrak{X}}$, so the χ-centre is a generalization of the $\mathfrak{X}$-radical. Of course, when χ is the property of being contained in the centre, $\zeta^\chi(G) = \zeta(G)$, the centre of G. By analogy with the upper central series we define the *upper χ-central series* of a group G,

$$1 = \zeta_0^\chi(G) \leqq \zeta_1^\chi(G) \leqq \cdots$$

by the rules

$$\zeta_{\alpha+1}^\chi(G)/\zeta_\alpha^\chi(G) = \zeta^\chi(G/\zeta_\alpha^\chi(G))$$

and

$$\zeta_\lambda^\chi(G) = \bigcup_{\beta<\lambda} \zeta_\beta^\chi(G)$$

where α is an ordinal and λ is a limit ordinal. The upper ζ-central series terminates with a subgroup called the *χ-hypercentre* of G, denoted by

$$\bar{\zeta}^\chi(G).$$

When χ is centrality, the χ-hypercentre is just the *hypercentre* or union of the upper central series

$$\bar{\zeta}(G).$$

When χ is the property of being a normal $\mathfrak{X}$-subgroup, the upper χ-central series is the series of successive $\mathfrak{X}$-radicals or *upper $\mathfrak{X}$-series.*

The following characterization of the χ-hypercentre is well-known.

Lemma 1.39.3. Let χ be a subgroup theoretical property which is inherited by homomorphic images. Then the χ-hypercentre of G is the intersection of all normal subgroups N of G such that $\zeta^\chi(G/N)$ is trivial.

Proof. By definition $G/\bar{\zeta}^\chi(G)$ has trivial χ-centre. Suppose that G/N has trivial χ-centre: we have to show that $\bar{\zeta}^\chi(G) \leqq N$ Assume that this is not the case. Then there is a first ordinal α such that

$$K_\alpha = \zeta_\alpha^\chi(G) \nleqq N.$$

Clearly α is not a limit ordinal, so $K_{\alpha-1} \leqq N$. By definition of K_α, there exists a subgroup M of G such that $(M/K_{\alpha-1})\, \chi(G/K_{\alpha-1})$ and $M \nleqq N$. Now χ is homomorphism invariant, so we may apply the natural homomorphism of $G/K_{\alpha-1}$ onto G/N, concluding that $(MN/N)\chi(G/N)$. But this implies that MN/N lies in the χ-centre of G/N, which is trivial. Thus $M \leqq N$, a contradiction. □

Finally, observe that *if χ is inherited by homomorphic images and if $\zeta^\chi(G)\, \chi G$ is always valid, then a group is a hyper-χ group if and only if it coincides with its χ-hypercentre.* This may be proved via Lemma 1.22 or directly.

The χ-Hypocentre

Let χ be a subgroup theoretical property and let G be a group. If N is a normal subgroup of G, we define

$$\gamma^\chi(N, G)$$

to be the intersection of all normal subgroups M of G such that $M \leq N$ and $(M/N)\ \chi(G/M)$ is valid. The terms of the *lower χ-central series* of G are defined by the rules

$$\gamma_1^\chi(G) = G, \quad \gamma_{\alpha+1}^\chi(G) = \gamma^\chi(\gamma_\alpha^\chi(G), G)$$

if α is an ordinal and

$$\gamma_\lambda^\chi(G) = \bigcap_{\beta<\lambda} \gamma_\beta^\chi(G)$$

if λ is a limit ordinal. The lower χ-central series of G terminates with the *χ-hypocentre* of G, denoted by

$$\bar{\gamma}^\chi(G).$$

When χ is centrality the lower χ-central series is simply the lower central series and $\bar{\gamma}^\chi(G)$ is the *hypocentre* of G. If χ is the property of being a normal $\mathfrak{X}$-subgroup for some class of groups $\mathfrak{X}$, the lower χ-central series is called the *lower $\mathfrak{X}$-series*. For example, taking $\mathfrak{X} = \mathfrak{A}$, we obtain the derived series.

Corresponding to Lemma 1.39.3 is the readily proven

Lemma 1.39.4. Let χ be a subgroup theoretical property satisfying the hypothesis of Lemma 1.24. Then the χ-hypocentre is the product of all normal subgroups N of G such that $\gamma^\chi(N, G) = N$.

In addition observe that if χ satisfies the hypothesis of Lemma 1.24 and if

$$(N/\gamma^\chi(N, G))\ \chi(G/\gamma^\chi(N, G))$$

is always valid, then a group is a hypo-χ group if and only if its χ-hypocentre is trivial. This follows via Lemma 1.24.

1.4 Finiteness Conditions

By a *finiteness condition* is meant any group theoretical property which is possessed by all finite groups. Naturally attention is directed primarily at finiteness conditions which are close to the property of being finite. In this section the most important finiteness conditions will be surveyed and some of the more elementary results about them obtained. Detailed discussion is reserved for Chapters Three, Four, Five and Nine.

A. Finitely Generated Groups

A group is said to be *finitely generated* if it has a finite subset which generates it; clearly, if this is the case, every generating set has a finite subset which also generates the group. A group with a set of n not necessarily distinct generators is called an *n-generator group*. The class

$$\mathfrak{G}$$

of finitely generated groups is obviously $\boldsymbol{H}$ and $\boldsymbol{P}$-closed. On the other hand, it is not $\boldsymbol{S}$ or even $\boldsymbol{S}_n$-closed: for example the group

$$\langle a, b : b^{-1}ab = a^2 \rangle \tag{14}$$

has a normal abelian subgroup, generated by $a, a^{b^{-1}}, a^{b^{-2}}, \ldots$, which is isomorphic with the additive group of all rational numbers whose denominators are powers of 2. Of course finitely generated groups, and therefore their subgroups, are countable, and a well-known theorem of Higman, Neumann and Neumann [1] asserts that the converse is true: in fact, any countable group G can be embedded in a 2-generator group G^*. Another proof of this result has been given by Levin [2].

Subsequently Neumann and Neumann showed that if G belongs to a variety $\mathfrak{V}$, then G^* can be taken to belong to the variety $\mathfrak{V}\mathfrak{A}^2$ ([2], Theorem 5.1).

P. Hall, generalizing earlier results of Dark [1] and Levin [5], has proved the following theorem (unpublished): *let H and K be non-unit groups, not both of order 2, and let G be any group such that*

$$|G| \leqq \max\{|H|, |K|, \aleph_0\};$$

then there is a group $J_1 = \langle H_1, K_1 \rangle$ where $H \simeq H_1$ and $K \simeq K_1$ with a subgroup G_1 such that $G \simeq G_1 \lhd^2 J_1$.

However, the situation is quite different for subgroups of finite index: this is the content of the fundamental

Theorem 1.41. If G is a finitely generated group and H is a subgroup with finite index in G, then H is finitely generated.

Proof. Let X be a finite set of generators of G and let $\{t_1, \ldots, t_i\}$ be a *right transversal* to H in G, that is a complete set of right coset representatives for H; we shall assume that $t_1 = 1$. If $g \in G$, then $Ht_jg = Ht_{(j)g}$ and

$$t_jg = h(j, g)\, t_{(j)g} \tag{15}$$

where $h(j, g) \in H$ and the mapping $j \to (j)\, g$ is a permutation of the set $\{1, \ldots, i\}$. Let $h \in H$ and write

$$h = y_1 \cdots y_k$$

where $y_j \in X \cup X^{-1}$. By repeated application of (15) we obtain

$$h = t_1 h = h(1, y_1)\, h((1) y_1, y_2) \cdots h((1) y_1 \cdots y_{k-1}, y_k)\, t_{(1)h}.$$

But $Ht_{(1)h} = Ht_1 h = H$, so $t_{(1)h} = 1$. It follows that

$$H = \langle h(j, y) : j = 1, \ldots, i, y \in X \cup X^{-1} \rangle.$$

Hence H is finitely generated. ☐

If G is an n-generator group and $|G : H| = i$, the well-known Reidemeister-Schreier Theorem asserts that H can be generated by

$$(n - 1)\, i + 1$$

elements. The proof of this depends on the theory of free groups (see for example Kuroš [9], § 36).

Next we record a result of Gregorac [1].

Theorem 1.42. If N is a normal subgroup of the finitely generated group G, if G/N is infinite cyclic and if $C_G(N) \nleq N$, then N is finitely generated.

Proof. Let $C = C_G(N)$: by hypothesis, there is an element x in C but not in N. Hence $H = \langle x, N \rangle \neq N$ and, since G/N is infinite cyclic, H has finite index in G. Therefore H is finitely generated, by Theorem 1.41. Since H/N is torsion-free, $\langle x \rangle \cap N = 1$ and $H = \langle x \rangle \times N$. Hence N is finitely generated. ☐

Example (14) shows that the condition $C_G(N) \nleq N$ cannot be omitted from Theorem 1.42.

B. Finitely Presented Groups

A *presentation* of group G is an exact sequence of groups

$$1 \to R \to F \xrightarrow{\theta} G \to 1 \tag{16}$$

where F is a free group. If G is finitely generated, F may be assumed to have finite rank. Let $\{x_1, \ldots, x_m\}$ be a set of free generators for F and let $x_i^\theta = a_i$, so that $G = \langle a_1, \ldots, a_m \rangle$. The exact sequence (16) is called a *finite presentation* of G if there exist finitely many elements $\varrho_1, \ldots, \varrho_n$ of F such that

$$\operatorname{Ker} \theta = \varrho_1^F \cdots \varrho_n^F.$$

G is said to be *finitely presented* by the generators $a_1, \ldots, a_m$ subject to the relations $\varrho_1 = \varrho_2 = \cdots = \varrho_n = 1$.

If the notion of a finitely presented group is to be useful, it must be shown that *the definition is independent of the presentation* (B. H. Neumann [2]). Let $b_1, \ldots, b_l$ be another finite set of generators for G: it will be shown that G can be presented by the b_i subject to a set of $l + n$ relations. For suitable words θ_i and ϕ_j we can write

$$a_i = \theta_i(b) \text{ and } b_j = \phi_j(a).$$

Then the relations

$$\varrho_k(\theta_1(b), \ldots, \theta_m(b)) = 1 \text{ and } b_j = \phi_j(\theta_1(b), \ldots, \theta_m(b)), \tag{17}$$

where $k = 1, \ldots, n$ and $j = 1, \ldots, l$, are certainly valid. Let $\bar{G}$ be the group presented by a set $\{\bar{b}_1, \ldots, \bar{b}_l\}$ subject to the $l + n$ relations (17) in the $\bar{b}_j$. Since all the defining relations of $\bar{G}$ are satisfied by the b_j, the map $\bar{b}_j \to b_j$ determines a homomorphism β of $\bar{G}$ onto G. Let $\bar{a}_i = \theta_i(\bar{b})$: then $\bar{G}$ can also be generated by $\bar{a}_1, \ldots, \bar{a}_m$ since $\bar{b}_j = \phi_j(\bar{a}_1, \ldots, \bar{a}_m)$. Since the original defining relations for G in the a_i are also valid in the $\bar{a}_i$, the map $a_i \to \bar{a}_i$ determines a homomorphism α of G onto $\bar{G}$. Now $\alpha\beta$ and $\beta\alpha$ are identity maps, so α and β are isomorphisms and $G \simeq \bar{G}$. Hence G is presented by the b_j subject to the relations (17).

In general neither subgroups nor homomorphic images of finitely presented groups are finitely presented. This is because of example (14) above and because free groups of finite rank are finitely presented. We mention an interesting theorem of Higman [7]: *every countable group with a recursively enumerable set of defining relations can be embedded in a finitely presented group.* For example, every countable abelian group and every countable locally finite group admit of such an embedding. Higman's theorem has been generalized by Clapham in [1] and [2]: see also Valiev [1] and McCool [2]. Also, in [1] Higman, Neumann and Neumann showed that *every countable group with a finite number of defining relations on some set of generators can be embedded in a group with 2-generators subject to the same number of relations.*

This lack of $\boldsymbol{S}$- and $\boldsymbol{H}$-closure makes the class of finitely presented groups difficult to deal with. However, the class is $\boldsymbol{P}$-closed, and there is a weak analogue of Theorem 1.41 available.

Lemma 1.43. (P. Hall [4], pp. 421 and 426). (i) If G is finitely generated and G/N is finitely presented, then N is finitely generated as a G-operator group.

(ii) The class of finitely presented groups is closed with respect to forming extensions.

Proof. (i) Let G be generated by $a_1, \ldots, a_m$ and let F be the free group on a set $\{x_1, \ldots, x_m\}$. The mapping $x_i \to a_i$ determines a homomorphism of F onto G: let R denote the inverse image of N under this homomorphism. The mapping $x_i \to a_iN$ determines a homomorphism of F onto G/N with kernel R, so

$$1 \to R \to F \to G/N \to 1$$

is a finite presentation of G/N. Hence $R = \varrho_1^F \cdots \varrho_n^F$ for some $\varrho_i = \varrho_i(x) \in F$ and $N = b_1^G \cdots b_n^G$ where $b_i = \varrho_i(a)$.

(ii) Let N be a normal subgroup of a group G and suppose that N is generated by $a_1, \ldots, a_m$ subject to relations $\varrho_1 = \cdots = \varrho_n = 1$ and that G/N is generated by $b_1N, \ldots, b_rN$ subject to relations $\sigma_1 = \cdots = \sigma_s = 1$. Evidently G can be generated by the elements

$$a_1, \ldots, a_m, b_1, \ldots, b_r. \tag{18}$$

For certain words λ_j, μ_{ij} and ν_{ij} there are relations $\sigma_i(b) = \lambda_i(a)$, $a_i^{b_j} = \mu_{ij}(a)$ and $a_i^{b_j^{-1}} = \nu_{ij}(a)$. Hence the following relations are satisfied by the generators (18),

$$\left.\begin{aligned} \varrho_i(a) &= 1, \ (i = 1, \ldots, n) \\ \sigma_j(b) &= \lambda_j(a), \ (j = 1, \ldots, s) \\ a_i^{b_j} = \mu_{ij}(a) \text{ and } a_i^{b_j^{-1}} &= \nu_{ij}(a) \quad (i = 1, \ldots, m, j = 1, \ldots, r). \end{aligned}\right\} \tag{19}$$

Let $\bar{G}$ be the group presented by generators $\bar{a}_1, \ldots, \bar{a}_m, \bar{b}_1, \ldots, \bar{b}_r$, subject to the relations (19) in the $\bar{a}_i$ and $\bar{b}_j$. It will be enough to prove that $G \simeq \bar{G}$. The mapping $\bar{a}_i \to a_i$, $\bar{b}_j \to b_j$ determines a homomorphism of $\bar{G}$ onto G: let K be the kernel of this homomorphism. Clearly $\bar{a}_i \to a_i$ determines an isomorphism of $\bar{N} = \langle \bar{a}_1, \ldots, \bar{a}_m \rangle$ with N, so $K \cap \bar{N} = 1$. Now $\bar{N} \lhd \bar{G}$ by the third set of defining relations in (19), so $\bar{b}_j\bar{N} \to b_jN$ determines an isomorphism of $\bar{G}/\bar{N}$ with G/N. Therefore $K \leq \bar{N}$ and $K = 1$. $\square$

Corollary. Every poly-(cyclic or finite) group is finitely presented.

In fact the poly-(cyclic or finite) groups are just the finite extensions of polycyclic groups, as elementary arguments show (see p. 65).

Wreath products of finitely presented groups (in their regular representations) have been studied by G. Baumslag [9]: he shows that *if H and K are finitely presented, the wreath product $H \wr K$ is finitely presented if and only if either H is a unit group or K is finite.*

The finitely presented groups which have received most attention are those with a single defining relation. An account of these groups may be found in the book by Magnus, Karrass and Solitar ([1], Section 4.4) and in the survey article of G. Baumslag [20].

C. Groups with Finite Rank

Following Prüfer, we will say that a group has *finite rank* r if every finitely generated subgroup can be generated by r elements and if r is the least positive integer with this property. If there is no such integer r, the group has rank ∞. The groups of rank 1 are just the locally cyclic groups and these are well-known to be the groups that are isomorphic with a subgroup of either the additive group of rational numbers or the multiplicative group of complex roots of unity. An abelian group A has finite rank r if and only if

$$r = r_0 + \max_p r_p < \infty$$

where the p-component of A is a direct product of r_p cyclic or quasicyclic groups $\neq 1$ and the factor group of A with respect to its torsion-subgroup is isomorphic with a subgroup of a direct product of r_0 but no fewer copies of the additive group of rational numbers. For a discussion of the different ranks of an abelian group the reader is referred to Fuchs [3], § 8.

Mal'cev [3] defines the *general rank* of a group G to be either ∞ or the least positive integer R such that every finitely generated subgroup is contained in an R-generator subgroup of G. (Mal'cev calls the rank of a group in the sense of Prüfer the *special rank*.) Clearly every group with finite rank r has finite general rank $R \leqq r$. If F is a free group on a set of n elements ($n \geqq 1$), then the general rank of F is n and the rank is ∞. Thus a "free group of rank n" is really a free group of general rank n.

In his paper [3] Mal'cev shows that there exist uncountable groups with finite general rank but that such groups cannot be linear or locally free. It seems to be an open question whether groups of finite rank are countable. Recently Platonov [6] has proved that a linear group of finite rank is soluble-by-finite.

Lemma 1.44 (Baer [51], Lemma 1.3). The class of groups of finite rank is ***S***, ***H*** and ***P***-closed.

Proof. Let $N \lhd G$ and suppose that N has finite rank r and G/N has finite rank r'. Let $H = \langle h_1, \ldots, h_m \rangle$ be a finitely generated subgroup of G. Then $h_i = \phi_i(a_1, \ldots, a_{r'})\, n_i$ for certain words ϕ_i and certain elements a_j of H and n_i of $H \cap N$. Also $\langle n_1, \ldots, n_m \rangle = \langle b_1, \ldots, b_r \rangle$ for some elements b_k of $H \cap N$. Hence $H = \langle a_1, \ldots, a_{r'}, b_1, \ldots, b_r \rangle$ and G has rank at most $r + r'$. Of course ***S*** and ***H***-closure are immediate. □

D. Periodic Groups, Groups of Finite Exponent, Locally Finite Groups

A *periodic* or *torsion group* is a group each of whose elements has finite order. The class of periodic groups

$$\mathfrak{P}$$

is evidently ***S***, ***H***, ***P***, ***N*** and ***L***-closed and is, in particular, a radical class. If G is a periodic group, the *exponent* of G is the least common multiple of the orders of the elements of G or, if this does not exist, ∞. The class of *locally finite groups*, groups in which every finitely generated subgroup is finite, is a subclass of the class of periodic groups. The problem whether every group of finite exponent, or more generally every periodic group, is locally finite is of course the celebrated *Burnside Problem.* The first example of a finitely generated infinite group with finite exponent was given recently by Novikov and Adjan [1]. Let F be a free group of rank m: Novikov and Adjan proved that *the Burnside group*

$$B(m, n) = F/F^n$$

is infinite if $m > 1$ and n is an odd integer $\geqq 4381$. This group has another remarkable property: *every non-trivial element has finite centralizer and in consequence every abelian subgroup is finite* (Novikov and Adjan [2]). In the other direction several authors have demonstrated that a group with exponent $e \leqq 4$ or $e = 6$ is locally finite: see Magnus, Karrass and Solitar [1], Sections 5.12 and 5.13.

The first published example of a finitely generated, infinite, periodic group is due to Golod [1]: the construction is based on the solution of a problem in class field theory by Golod and Šafarevič [1]. Golod's group is a residually finite p-group of infinite exponent.

In a series of papers [1], [2], [4] Kostrikin has obtained results in the theory of Lie algebras which lead to a positive solution of the so-called *restricted Burnside Problem* for prime exponents: *there is an upper bound $d(m, p)$ for the order of a finite group of prime exponent p with m generators.*

Generally speaking Burnside problems will play little part in this work: for a survey of results in this field the reader should consult Magnus, Karrass and Solitar [1], Chapter 5, or Kuroš [13], Vol. 3, Chapter 18.

Theorem 1.45 (O. J. Schmidt [5], Theorem 6). The class of locally finite groups is closed with respect to forming extensions.

Proof. Let $N \lhd G$ and suppose that both N and G/N are locally finite. If H is a finitely generated subgroup of G, then HN/N, and hence $H/H \cap N$, is finite. By Theorem 1.41 the subgroup $H \cap N$ is finitely generated, so it is finite. Consequently H is finite and G is locally finite. $\square$

Corollary. The class of locally finite groups is a $\acute{\boldsymbol{P}}$-closed radical class.

Proof. Since $\boldsymbol{N}_0 \leqq \boldsymbol{PH}$ and $\boldsymbol{L}\mathfrak{F}$ is obviously $\boldsymbol{H}$-closed, we deduce from the theorem that $\boldsymbol{L}\mathfrak{F}$ is $\boldsymbol{N}_0$-closed. $\boldsymbol{L}$-closure is trivial, so $\boldsymbol{L}\mathfrak{F}$ is $\boldsymbol{N}$-closed and Lemma 1.32 shows that $\boldsymbol{L}\mathfrak{F}$ is a radical class. The $\acute{\boldsymbol{P}}$-closure of $\boldsymbol{L}\mathfrak{F}$ follows from the easily established relation $\acute{\boldsymbol{P}} \leqq \langle \boldsymbol{L}, \boldsymbol{P} \rangle$. $\square$

Some Classes of Locally Finite Groups

We consider now some particular types of locally finite groups. In [2] Specht introduced the class $\mathfrak{X}$ (temporary notation) of all groups G which have a well-ordered ascending chain of subgroups

$$1 = G_0 < G_1 < \cdots < G_\beta < G_{\beta+1} \cdots G_\alpha = G$$

such that G_β has finite index in $G_{\beta+1}$ for each $\beta < \alpha$. A simple transfinite induction on α shows that G is locally finite. Successively narrower subclasses of $\mathfrak{X}$ are obtained by applying to $\mathfrak{F}$ the closure operations $\acute{\boldsymbol{P}}$, $\acute{\boldsymbol{P}}_{sn}$ and $\acute{\boldsymbol{P}}_n$. By Theorem 1.33 we see that $\acute{\boldsymbol{P}}_{sn}\mathfrak{F}$ is the radical class generated by $\mathfrak{F}$, while $\acute{\boldsymbol{P}}_n\mathfrak{F}$ is the class of *hyperfinite* groups.

If $\mathfrak{Y}$ is any class of groups, we define

$$\boldsymbol{M}\mathfrak{Y},$$

the class of *locally normal and $\mathfrak{Y}$-groups*, to consist of all groups G in which each finite subset lies in a normal $\mathfrak{Y}$-subgroup. Obviously $\boldsymbol{M}$ is an operation but not a closure operation. The class $\boldsymbol{M}\mathfrak{F}$ of locally normal and finite groups is a subclass of $\acute{\boldsymbol{P}}_n\mathfrak{F}$ which has been much studied.

The inclusion relations between these classes of groups are

$$\boldsymbol{M}\mathfrak{F} < \acute{\boldsymbol{P}}_n\mathfrak{F} < \acute{\boldsymbol{P}}_{sn}\mathfrak{F} < \acute{\boldsymbol{P}}\mathfrak{F} < \mathfrak{X} < \boldsymbol{L}\mathfrak{F}.$$

The strictness of the inclusions is demonstrated by five examples, of which three must wait till a later chapter for their construction.

(a) $\mathfrak{X} < \boldsymbol{L}\mathfrak{F}$: the Kargapolov-Kovács-Neumann example (Part 2, p. 28).

(b) $\acute{\boldsymbol{P}}\mathfrak{F} < \mathfrak{X}$: for example, the alternating group on a countably infinite set is the union of an ascending chain of finite simple groups, so it belongs to $\mathfrak{X}$ and yet it contains no non-trivial ascendant finite subgroups.

(c) $\acute{\boldsymbol{P}}_{sn}\mathfrak{F} < \acute{\boldsymbol{P}}\mathfrak{F}$: the wreath power $Wr\, H^{Z^-}$ where H is a group of prime order p and Z^- is the set of negative integers in their natural order (Part 2, p. 22).

(d) $\acute{\boldsymbol{P}}_n\mathfrak{F} < \acute{\boldsymbol{P}}_{sn}\mathfrak{F}$: Čarin's example (p. 152).

(e) $\boldsymbol{M}\mathfrak{F} < \acute{\boldsymbol{P}}_n\mathfrak{F}$: for example the so-called *locally dihedral 2-group*, that is the holomorph of a group of type 2^∞ by the automorphism group generated by the automorphism $a \to a^{-1}$.

Finally we mention groups which have a *finite category*. In general, if $\mathfrak{Y}$ is any class of groups and α is an ordinal, a class of group $\mathfrak{Y}^{(\alpha)}$ is defined as follows: $\mathfrak{Y}^{(1)} = \mathfrak{Y}$ and $G \in \mathfrak{Y}^{(\alpha)}$, $\alpha > 1$, if and only if G is the set theoretical union of normal subgroups each of which belongs to a class $\mathfrak{Y}^{(\beta)}$ where $\beta < \alpha$.

The class of groups which have a $\mathfrak{Y}$-*category* is defined to be

$$\text{Cat}\ \mathfrak{Y} = \bigcup_{\alpha} \mathfrak{Y}^{(\alpha)},$$

the union being formed over all ordinals α. The class Cat $\mathfrak{A}$ was first introduced by Kontorovič [7].

Thus $\mathfrak{F}^{(2)}$ is the class of locally normal and finite groups. Kalužnin [4] has proved that $\mathfrak{F}^{(i)} < \mathfrak{F}^{(i+1)}$ for each integer i. We shall see later (in the Corollary to Theorem 2.36) that

$$\text{Cat}\ \mathfrak{F} = \boldsymbol{N}\mathfrak{F}.$$

Notice that $\boldsymbol{N}\mathfrak{F} \leq \acute{\boldsymbol{P}}_{sn}\mathfrak{F}$ since the latter is a radical class.

E. Maximal and Minimal Conditions

Some of the most important finiteness conditions in the theory of groups are *chain conditions* which exclude the possibility of infinite well-ordered ascending or descending chains of subgroups of certain types.

Let f be a function assigning to a group G a set $f(G)$ of subgroups of G such that if θ is an isomorphism of G with some other group, then

$$f(G^\theta) = \{H^\theta : H \in f(G)\}.$$

f is called a *group theoretical functional**. Notice that when $f(G)$ consists of just one subgroup f is a group theoretical function as defined on *p*. 18.

A group G satisfies the *ascending chain condition on f-subgroups* if there are no infinite strictly-ascending chains in $f(G)$ and G satisfies the *descending chain condition on f-subgroups* if there are no infinite strictly-descending chains in $f(G)$. The following result is well-known and simple to prove.

Lemma 1.46. If f is a group theoretical functional, a group G satisfies the ascending (descending) chain condition on f-subgroups if and only if each non-empty subset of $f(G)$ has at least one maximal (minimal) element.

Following general usage we will call the properties of Lemma 1.46 *the maximal and minimal conditions on f-subgroups* and denote them by

$$\text{Max-}f \text{ and Min-}f.$$

* It may be that $f(G)$ is defined only if G belongs to some particular class of groups.

The Most Important Cases

(i) Let Σ be a set of multiplicative operators on a group G, so that G is a Σ-operator group (see Kuroš [9], § 15). Let $\mathfrak{f}(G)$ be the set of all Σ-admissible subgroups of G. Then we obtain the finiteness conditions

$$\text{Max-}\Sigma \text{ and Min-}\Sigma.$$

There is a well-known criterion for a Σ-operator group to satisfy Max-Σ.

Lemma 1.47. A Σ-operator group G satisfies Max-Σ if and only if each Σ-admissible subgroup of G can be finitely generated as a Σ-operator group.

When $\Sigma = \text{End } G$, Aut G, Inn G or the identity automorphism group, Max-Σ and Min-Σ become the maximal and minimal conditions on fully invariant subgroups, characteristic subgroups, normal subgroups or subgroups respectively. The last two cases are the most useful:

$$\text{Max-}n \text{ and Min-}n\text{; Max and Min.*}$$

(ii) $\mathfrak{f}(G)$ = the set of all subnormal subgroups of G. The *maximal and minimal conditions on subnormal subgroups* are denoted by

$$\text{Max-}sn \text{ and Min-}sn.$$

(iii) $\mathfrak{f}(G)$ = the set of all abelian subgroups of G. The properties

$$\text{Max-}ab \text{ and Min-}ab$$

are equivalent to every abelian subgroup being finitely generated and to every abelian subgroup satisfying Min respectively.

In the following lemma Σ is a set of multiplicative operators and $\mathfrak{f}$ is a group theoretical functional assigning to each Σ-operator group G a set $\mathfrak{f}(G)$ of Σ-admissible subgroups of G.

Lemma 1.48. Consider the following conditions on $\mathfrak{f}$.

(a) If θ is a Σ-operator homomorphism of a group G and $X \in \mathfrak{f}(G)$, then $X^\theta \in \mathfrak{f}(G^\theta)$.

(b) If N is a normal Σ-admissible subgroup of a Σ-operator group G and $X \in \mathfrak{f}(G)$, then $X \cap N \in \mathfrak{f}(N)$.

(c) If N is a normal Σ-admissible subgroup of a Σ-operator group G and $X \in \mathfrak{f}(G)$, then $X \cap N \in \mathfrak{f}(G)$.

If $\mathfrak{f}$ satisfies (a) and (b), then Max-$\mathfrak{f}$ and Min-$\mathfrak{f}$ are $\boldsymbol{P}$-closed and if $\mathfrak{f}$ satisfies (a) and (c), then Max-$\mathfrak{f}$ and Min-$\mathfrak{f}$ are $\boldsymbol{R}_0$-closed.

---

* Groups with Max are often called *noetherian* and groups with Min *artinian*.

Proof. Let $\mathfrak{f}$ satisfy (a) and (c). Let M and N be two normal Σ-admissible subgroups of a Σ-operator group G such that $M \cap N = 1$. To establish the $\boldsymbol{R}_0$-closure of Max-$\mathfrak{f}$ it is enough to show that G has Max-$\mathfrak{f}$ if G/M and G/N have Max-$\mathfrak{f}$. If this is not the case, there is a countably infinite, ascending chain $X_1 < X_2 < \cdots$ in $\mathfrak{f}(G)$. Now $X_iM/M \in \mathfrak{f}(G/M)$ by (a); also $X_i \cap M \in \mathfrak{f}(G)$ by (c), and by (a) once again we obtain $(X_i \cap M)\, N/N \in \mathfrak{f}(G/N)$. In view of Max-$\mathfrak{f}$, there is a positive integer i such that $X_iM = X_{i+1}M =$ etc. and

$$(X_i \cap M)\, N = (X_{i+1} \cap M)\, N = \text{etc.}$$

By Dedekind's modular law

$$X_{i+1} = X_{i+1} \cap (X_iM) = X_i(X_{i+1} \cap M)$$

and

$$X_{i+1} \cap M = (X_{i+1} \cap M) \cap ((X_i \cap M)\, N) = X_i \cap M$$

since $M \cap N = 1$. But these equations lead to the contradiction $X_i = X_{i+1}$.

A similar argument deals with Min-$\mathfrak{f}$ and $\boldsymbol{P}$-closure. ▯

Corollary. The following properties are both $\boldsymbol{P}$ and $\boldsymbol{R}_0$-closed:

Max, Max-*sn*, Max-*ab*, Max-Σ, Min, Min-*sn*, Min-*ab*, Min-Σ.

F. Linear Groups

Let n be a positive integer and let F be a field. A group G is said to be *linear of degree n over F* it if is isomorphic with a subgroup of $GL(n, F)$, the group of all $n \times n$ non-singular matrices over F or, equivalently, if it is isomorphic with a group of invertible linear transformations of a vector space of dimension n over F. Linearity is a finiteness condition since every finite group has a faithful representation as a group of permutations, for example, the right regular representation. On the other hand, not every infinite group is linear (see p. 79). Thus the behaviour of linear groups may be expected to be better than that of arbitrary groups. For example, a classical result of Schur [4] asserts that a periodic linear group over the field of complex numbers is locally finite (see also Curtis and Reiner [1], § 36): this has been extended to periodic linear groups over an arbitrary field, and even over an arbitrary commutative ring with unity, by Tokarenko [1]. Thus the Burnside Problem has a positive solution for linear groups.

The fundamental paper of Mal'cev [1] deals with group theoretical properties of linear groups. There it is shown that *every finitely generated linear group of degree n is residually* (*finite and linear of degree n*) (Theorems VII and VIII), and also that *the class of linear groups of degree n is*

L-closed (Theorem IV).* Now it was shown by Brenner [1] that the matrices

$$\begin{pmatrix} 1 & m \\ 0 & 1 \end{pmatrix} \text{ and } \begin{pmatrix} 1 & 0 \\ m & 1 \end{pmatrix}$$

generate a free group of rank 2 for all real $m \geqq 2$—there is a generalization in a paper of M. Newman [1]: see also Lyndon and Ullman [1]. Hence, by the **S**- and **L**-closure of the class of linear groups of degree 2, *every free group is linear of degree* 2. Consequently, the class of linear groups is not **H**-closed.

Finally, we mention that linear groups of infinite degree are studied in a paper of Plotkin [25].

G. Hopfian groups

A group G is called *hopfian* (after H. Hopf) if G is not isomorphic with some factor group G/N where $N \neq 1$: this is equivalent to requiring that every endomorphism of G which is an epimorphism be an automorphism. Clearly, every finite group is hopfian, so hopficity is a finiteness condition. More generally a group G which satisfies Max-n, the maximal condition on normal subgroups, is hopfian: for if G is not hopfian, $G \simeq G/N_1$ for some $N_1 > 1$ and $G/N_1 \simeq G/N_2$ for some $N_2 > N_1 > 1$ and so on: in this way we produce an infinite ascending chain of normal subgroups of G.

Originally Hopf asked whether every finitely generated group is hopfian. This was answered in the negative by B. H. Neumann [6], and shortly afterwards Higman [1] gave an example of a finitely presented non-Hopfian group. Since then many other examples of finitely generated non-hopfian groups have been given with additional properties: see Neumann and Neumann [2] (Theorem 7.1), P. Hall [11] (Theorem 4), Baumslag and Solitar [1], G. Baumslag [10] and M. F. Newman [4]. Also G. Baumslag [13] has constructed for each cardinal $\mathfrak{c}$ an abelian hopfian group of cardinality $\mathfrak{c}$.

Theorem 1.49. Let G be a residually finite group such that there exist at most a finite number of distinct homomorphisms of G onto a given finite group. Then G is hopfian.

Proof. Suppose that $G \simeq G/N$ where $N \neq 1$ and choose a non-trivial element x from N. Since G is residually finite, there is a normal subgroup M of finite index in G to which x does not belong. Let $H = G/M$. By hypothesis there is only a finite number—say n—of homomorphisms of

* For another proof see Kegel [8], p. 75. A special case is in Nisnevič [1].

G onto H. By mapping G onto G/N by means of the natural homomorphism and then G/N onto H we obtain n distinct homomorphisms of G onto H such that x belongs to the kernel in each case. However, the natural homomorphism of G onto H does not have x in its kernel, so there are $n + 1$ distinct homomorphisms of G onto H, a contradiction. $\square$

Corollary. Finitely generated residually finite groups are hopfian.

Since Mal'cev has proved that finitely generated linear groups are residually finite, it follows that *finitely generated linear groups are hopfian* (Mal'cev [1] Theorem XII). In particular, *free groups of finite rank are hopfian* (Levi [1]): this also follows from Iwasawa's Theorem that free groups are residually finite (see Theorem 9.11 below). Therefore the class of hopfian groups is neither $\boldsymbol{S}$ nor $\boldsymbol{H}$-closed.

It is easy to prove that a group with an ascending hopfian series each of whose terms is mapped onto itself by every endomorphism which is also an epimorphism is hopfian: this and similar results are in a paper of Baer [10].

Finally we remark that the dual property, co-hopficity, has been studied by Baer in [10] and [53]: a group is called *co-hopfian* if it is not isomorphic with a proper subgroup.

Chapter 2

Soluble and Nilpotent Groups

2.1 Conjugates and Commutators. Some Generalized Soluble and Nilpotent Groups

The calculus of commutators is fundamental to this study; in this section we will establish our notation and some basic properties.

Let $x, y, z, \dots$ be elements of a group. Then, of course, the *conjugate of x by y* is

$$x^y = y^{-1}xy$$

and the *commutator of x with y* is

$$[x, y] = x^{-1}y^{-1}xy = x^{-1}x^y.$$

More generally, simple commutators of weight $n + 1$ are defined inductively by the rules

$$[x_1, \dots, x_{n+1}] = [[x_1, \dots, x_n], x_{n+1}] \text{ and } [x_1] = x_1.$$

As is well-known, the identities

$$[x, yz] = [x, z]\,[x, y]^z \tag{1}$$

$$[xy, z] = [x, z]^y\,[y, z] \tag{2}$$

and

$$[x, y^{-1}, z]^y\,[y, z^{-1}, x]^z\,[z, x^{-1}, y]^x = 1 \tag{3}$$

are valid in any group. Verification of (1) and (2) is immediate, while (3) is most readily proved by setting $u = xzx^{-1}yx$, $v = yxy^{-1}zy$ and $w = zyz^{-1}xz$ and observing that $u^{-1}v = [x, y^{-1}, z]^y$, $v^{-1}w = [y, z^{-1}, x]^z$ and $w^{-1}u = [z, x^{-1}, y]^x$: the truth of (3) is now evident.

It is important to be able to form conjugates and commutators of subsets as well as elements of a group and these are defined in the following way. Let $X, Y, Z, \dots$ be non-empty subsets of a group. The conjugate of X by Y or, as it is usually called, *the normal closure of X in Y*, is the *subgroup*

$$X^Y = \langle x^y : x \in X, y \in Y \rangle$$

generated by all conjugates of elements of X by elements of Y. If Y is a subgroup, X^Y is clearly normal in $\langle X, Y\rangle$, and indeed it is the smallest normal subgroup of $\langle X, Y\rangle$ that contains X, which accounts for the term "normal closure". It follows that, whenever Y is a subgroup,

$$X^Y = X^{\langle X,Y\rangle}.$$

Closely related to X^Y is the *commutator subgroup of X with Y*

$$[X, Y] = \langle [x, y] : x \in X, y \in Y\rangle.$$

Since $[x, y] = [y, x]^{-1}$, we have $[X, Y] = [Y, X]$. We define inductively

$$[X_1] = \langle X_1\rangle \text{ and } [X_1, \ldots, X_{n+1}] = [[X_1, \ldots, X_n], X_{n+1}], \quad (n > 1).$$

If A is a group of automorphisms of a group G, symbols like $[g, \alpha]$, $(g \in G, \alpha \in A)$, and $[G, A]$ are to be interpreted according to the convention that G and A are subgroups of the holomorph of G by A.

Lemma 2.11. Let X and Y be non-empty subsets of a group.

(i) If $1 \in Y$, then $X^Y = \langle X, [X, Y]\rangle$.

(ii) If Y is a subgroup, $[X, Y]^Y = [X, Y]$: consequently $[X, Y, Y] \leq [X, Y]$.

Proof. (i) follows at once from the definitions. To prove (ii) let $x \in X$ and let $y_1, y_2 \in Y$. Then by equation (1)

$$[x, y_1]^{y_2} = [x, y_2]^{-1}[x, y_1y_2] \in [X, Y]. \quad \square$$

Two special cases are worth mentioning. Let H and K be subgroups of a group and let x be an element of the group. Then

$$[H, K] \lhd \langle H, K\rangle$$

and

$$H^K = H[H, K] = H^{\langle H,K\rangle},$$

while

$$[H, x] \lhd \langle H, [H, x]\rangle = \langle H, H^x\rangle.$$

Lemma 2.12. Let X and Y be non-empty subsets of a group and let $H = \langle X\rangle$. Then

$$[H, Y] = [X, Y]^H.$$

Proof. $[X, Y]^H \leq [H, Y]^H = [H, Y]$ by Lemma 2.11. The opposite inclusion follows from the equations

$$[x_1^{-1}, y] = ([x_1, y]^{-1})^{x_1^{-1}}$$

and

$$[x_1^{\varepsilon_1} \cdots x_{k+1}^{\varepsilon_{k+1}}, y] = [x_1^{\varepsilon_1} \cdots x_k^{\varepsilon_k}, y]^{x_{k+1}^{\varepsilon_{k+1}}} [x_{k+1}^{\varepsilon_{k+1}}, y],$$

where $x_i \in X$, $y \in Y$ and $\varepsilon_i = \pm 1$, and an induction on k. □

Corollary. Let H and K be subgroups of a group and suppose that $H = \langle X \rangle$ and $K = \langle Y \rangle$. Then

$$[H, K] = ([X, Y]^H)^K = [X, Y]^{HK}.$$

Lemma 2.13. (*The Three Subgroup Lemma*). Let H, K and L be subgroups of a group G. If any two of the subgroups $[H, K, L]$, $[K, L, H]$ and $[L, H, K]$ are contained in a normal subgroup of G, then so is the third.

A special case of this result appeared first in the fundamental paper of P. Hall [1] (Theorem 2.3): for the above version see Kalužnin [1], [3] and P. Hall [7] (Lemma 1). To prove Lemma 2.13 observe that $[H, K, L]$ may be generated by conjugates of elements of the form $[h, k, l]$, where $h \in H$, $k \in K$ and $l \in L$, and apply equation (3).

Normal closures and products of conjugates

A normal closure is always generated by conjugates, but when is it a *product* of conjugates and, if it is, in what order can the conjugates occur? The answer is furnished by the following simple principle.

Lemma 2.14. Let $\mathscr{S}$ be a set of subgroups of a group and suppose that $\mathscr{S}$ has the property that if X and Y belong to $\mathscr{S}$, so does X^y where $y \in Y$. Let $\mathscr{S}$ be well-ordered in any manner. Then the subgroup J generated by all the members of $\mathscr{S}$ is just the product of the members of $\mathscr{S}$ in the prescribed order.

Proof. Let $\mathscr{S}$ be well-ordered as $\{X_\alpha : \alpha < \beta\}$ where β is an ordinal number and let $1 \neq a \in J$. Then certainly a has the form $x_1 \cdots x_r$ where $x_i \in X_{\alpha_i}$ and $\alpha_i < \beta$. Now among all such expressions for a there are some for which the length r is minimal and among the latter there is one expression which occurs first in the lexicographic ordering of r-tuples: this is the ordering in which $(\alpha_1, \ldots, \alpha_r)$ precedes $(\alpha_1', \ldots, \alpha_r')$ if $\alpha_t = \alpha_t'$ for $t < i$ but $\alpha_i < \alpha_i'$ for some $i \leq r$. Let $a = x_1 \cdots x_r$ be this expression for a. Assume that $\alpha_i = \alpha_j$ for some $i \neq j$. If, say, $i < j$, we can write

$$a = x_1 \cdots x_{i-1} (x_i x_j) x_{i+1}^{x_j} \cdots x_{j-1}^{x_j} x_{j+1} \cdots x_r$$

by moving x_j to the ith position. Since $x_i x_j \in X_{\alpha_i}$ and $x_k^{x_j} \in X_k^{x_j} \in \mathscr{S}$ by hypothesis, this is an expression for a of length less than r, which is impossible. Next suppose that $\alpha_{i+1} < \alpha_i$: then

$$a = x_1 \cdots x_{i-1} x_{i+1} x_i^{x_{i+1}} x_{i+2} \cdots x_r.$$

But this is an expression for a which precedes the chosen one in the lexicographic ordering. Hence in the expression $a = x_1 \cdots x_r$ we must have $x_i \in X_{\alpha_i}$ where $\alpha_1 < \alpha_2 < \cdots < \alpha_r < \beta$, so the lemma is proved. □

Corollary 1 (Szép and Itô [1]). *If H is a subgroup of a group G and if the set of conjugates of H in G is well-ordered in any manner, then H^G is the product of the conjugates of H in this order.*

The second application of Lemma 2.14 is an important result generally known as *Dietzmann's Lemma.*

Corollary 2 (Dietzmann [1]). In any group a normal finite subset consisting of elements of finite order generates a finite subgroup.

(A subset of a group is *normal* if it contains all conjugates of its elements).

Corollary 3. Let H be a subgroup of a group G. Then H^G is finite if and only if H is finite and $|G: N_G(H)|$ is finite.

This elementary finiteness criterion is a consequence of both Corollary 1 and Corollary 2.

Soluble and Generalized Soluble Groups

A group is said to be *soluble* (or solvable) if it has an abelian series of finite length. Thus the soluble groups are just the *polyabelian groups* and they constitute the class

$$\mathfrak{S} = P\mathfrak{A}.$$

The *derived series* of a group G is the descending fully-invariant series $\{G^{(\alpha)}\}$ defined by

$$G^{(0)} = G \text{ and } G^{(\alpha+1)} = [G^{(\alpha)}, G^{(\alpha)}] = (G^{(\alpha)})',$$

with the usual requirement $G^{(\lambda)} = \bigcap_{\beta<\lambda} G^{(\beta)}$ when λ is a limit ordinal. This series terminates eventually, that is to say $G^{(\beta)} = G^{(\beta+1)} =$ etc. for some first ordinal $\beta = \beta(G)$. Clearly G is soluble if and only if $G^{(d)} = 1$ for some finite d; the least integer d with this property is the *derived or soluble length* of G.

The concept of solubility lends itself to generalization and some examples have been mentioned already in Section 1.3. A very wide class of generalized soluble groups* is the class

$$\hat{\boldsymbol{P}}\mathfrak{A}$$

of groups with an abelian series: such groups are usually called *SN-groups*. Obviously $\hat{\boldsymbol{P}}\mathfrak{A} = \hat{\boldsymbol{P}}\mathfrak{C}$. Much narrower is the class

$$\hat{\boldsymbol{P}}_n\mathfrak{A}$$

of groups with a normal abelian series or *SI-groups*: these, unlike *SN*-groups, cannot be simple and non-abelian (see Section 8.4 in this connection). By employing ascending and descending series we can obtain further subclasses of $\hat{\boldsymbol{P}}\mathfrak{A}$, namely the classes

$$\acute{\boldsymbol{P}}\mathfrak{A} \text{ and } \grave{\boldsymbol{P}}\mathfrak{A}$$

of *SN*-groups* and *hypoabelian groups* respectively. The latter are precisely the groups in which the derived series reaches the identity subgroup, so

$$\grave{\boldsymbol{P}}\mathfrak{A} = \grave{\boldsymbol{P}}_n\mathfrak{A};$$

the hypoabelian groups appeared in Section 1.3 as the residual class generated by the class of abelian groups. Another class which we have already met is the class

$$\acute{\boldsymbol{P}}_{sn}\mathfrak{A}$$

of *subsoluble groups* (or *SJ*-groups*): this subclass of $\acute{\boldsymbol{P}}\mathfrak{A}$ is the radical class generated by the class of abelian groups. Still smaller is the class

$$\acute{\boldsymbol{P}}_n\mathfrak{A}$$

of *hyperabelian groups* (or *SI*-groups*). Finally we mention the *locally soluble groups*, one of the most interesting types of generalized soluble groups.

Some other forms of generalized solubility will arise later. In Chapter 8 a detailed discussion of the relation between the various classes of generalized soluble groups can be found.

Hyperabelian Groups

By Lemma 1.22 *a group is hyperabelian if and only if each non-trivial homomorphic image has a non-trivial normal abelian subgroup.* A quite

* For the exact definition of the term *generalized soluble* see Section 6.1.

different characterization of hyperabelian groups has been devised by Baer in [20] (§ 4, Satz 1): this runs as follows.

Theorem 2.15. A group G is hyperabelian if and only if given two sequences $x_1, x_2, \ldots$ and $y_1, y_2, \ldots$ of elements of G such that

$$x_{i+1} = [x_i, y_i, x_i],$$

there is an integer $m > 0$ such that $x_m = 1$.

Proof. Suppose that G is hyperabelian and that

$$1 = G_0 \leqq G_1 \leqq \cdots \leqq G_\alpha = G$$

is an ascending normal abelian series in G. Let $x_1, x_2, \ldots$ and $y_1, y_2, \ldots$ be two sequences in G such that $x_{i+1} = [x_i, y_i, x_i]$ and assume that $x_i \neq 1$ for all i. Then $x_i \in G_{\alpha_i+1} \backslash G_{\alpha_i}$ for some ordinal α_i. Now, since $G_{\alpha_i+1} \lhd G$,

$$x_{i+1} = [x_i, y_i, x_i] \in (G_{\alpha_i+1})' \leqq G_{\alpha_i},$$

so $\alpha_{i+1} < \alpha_i$. Hence $\alpha_1 > \alpha_2 > \cdots$ is an infinite descending chain of ordinals, which of course cannot exist.

Conversely assume that G satisfies the condition on sequences. This condition is easily seen to be inherited by homomorphic images, so it will be enough to prove that if G is non-trivial, it contains a non-trivial normal abelian subgroup.

Suppose that this is not the case: if $1 \neq x_1 \in G$, then x_1^G is not abelian, so there is a $y_1 \in G$ such that $[x_1^{y_1}, x_1] \neq 1$. Let $x_2 = [x_1^{y_1}, x_1]$ and note that $x_2 = [x_1, y_1, x_1]$. Again x_2^G is not abelian, so there is a $y_2 \in G$ such that $x_3 = [x_2^{y_2}, x_2] = [x_2, y_2, x_2] \neq 1$. In this way we can generate two sequences of the type in question consisting of non-trivial elements, which is contrary to our hypothesis. ▯

Corollary (Baer [32], p. 360). A group is hyperabelian if and only if every countable subgroup is hyperabelian.

More general theorems of this type are obtained in Section 8.3. We remark that further characterizations of hyperabelian groups are contained in papers of Ščukin [4] and Baer [51] (Lemma 4.2).

Lemma 2.16. Let χ be a subgroup theoretical property which is inherited by homomorphic images and which satisfies the hypothesis of Lemma 1.24 If M is a non-trivial normal subgroup lying in the χ-hypercentre of a group G, then M contains a non-trivial normal subgroup N of G such that $N\chi G$.

Proof. Since χ is homomorphism closed, we see easily that the upper χ-central series of G can be refined to an ascending G-admissible series

$\{G_\alpha\}$ such that $(G_{\alpha+1}/G_\alpha)\,\chi(G/G_\alpha)$ for all α. Since M lies in the χ-hypercentre there is a first ordinal α for which $M \cap G_\alpha \neq 1$. Then α is not a limit ordinal and $M \cap G_{\alpha-1} = 1$. Now $(G_\alpha/G_{\alpha-1})\,\chi(G/G_{\alpha-1})$, so the hypothesis on χ leads us to $M \cap G_\alpha \chi G$. ☐

This applies in particular to non-trivial normal subgroups of hyper-$\mathfrak{X}$ groups where $\mathfrak{X} = \boldsymbol{S}_n\mathfrak{X} = \boldsymbol{H}\mathfrak{X}$. It also implies that *if M is a non-trivial normal subgroup lying in the hypercentre of a group G, then $M \cap \zeta(G) \neq 1$.*

We shall use Lemma 2.16 to establish

Lemma 2.17. Let G be a hyperabelian group and let F be its Fitting subgroup. Then $C_G(F) \leq F$.

Proof. Let $C = C_G(F)$ and suppose that $C \nleq F$. By Lemma 2.16 the group CF/F contains a non-trivial normal abelian subgroup A/F of G/F. Now $A \cap C \lhd G$ and $[(A \cap C)', A \cap C] \leq [F, C] = 1$, so $A \cap C$ is nilpotent; hence $A \cap C \leq F$. But $A = A \cap (CF) = (A \cap C)\,F = F$, which is not so. ☐

Nilpotent and Generalized Nilpotent Groups

We recall that a normal series $\{\Lambda_\sigma, V_\sigma\colon \sigma \in \Sigma\}$ in a group G is a *central series* if Λ_σ/V_σ lies in the centre of G/V_σ for all $\sigma \in \Sigma$. For any group G the *lower central series* of G is the descending central series $\{\gamma_\alpha(G)\}$ of fully invariant subgroups defined by the rules

$$\gamma_1(G) = G \text{ and } \gamma_{\alpha+1}(G) = [\gamma_\alpha(G), G].$$

Let the centre of G be denoted by

$$\zeta(G).$$

The *upper central series* of G is the ascending central series $\{\zeta_\alpha(G)\}$ of characteristic subgroups defined by

$$\zeta_0(G) = 1 \text{ and } \zeta_{\alpha+1}(G)/\zeta_\alpha(G) = \zeta(G/\zeta_\alpha(G)).$$

Thus $\zeta_1(G) = \zeta(G)$. Both the upper and lower central series of G terminate, so there exist least ordinals $\beta = \beta(G)$ and $\beta^* = \beta^*(G)$ such that $\zeta_\beta(G) = \zeta_{\beta+1}(G) =$ etc. and $\gamma_{\beta^*}(G) = \gamma_{\beta^*+1}(G) =$ etc.

If $\beta(G) \geq 2$, then $\beta^*(G) \geq 2$: for let $z \in \zeta_2(G)\backslash\zeta_1(G)$ and observe that the map $g \to [z, g]$ is a homomorphism of G into $\zeta(G)$ with non-trivial image; hence $G > G' = \gamma_2(G)$. This result is usually referred to as *Grün's Lemma* (Grün [1], § 2, Satz 4). However it is interesting that apart from this there is very little relation between the lengths of the upper and lower central series, as has been demonstrated by Meldrum [1]. The last term of the upper central series of G is the hypercentre,

$$\bar{\zeta}(G).$$

Most of these ideas have already appeared in Chapter 1 in the more general setting of subgroup theoretical properties.

A group G is said to be *nilpotent* if it has a central series of finite length. It follows easily from the definitions that a group G is nilpotent if and only if $\gamma_{c+1}(G) = 1$ for some integer $c \geqq 0$, and that $\gamma_{c+1}(G) = 1$ if and only if $\zeta_c(G) = G$. The least integer c such that $\gamma_{c+1}(G) = 1$ or, equivalently, such that $\zeta_c(G) = G$, is called the *nilpotent class* of G.

A fundamental fact about $\mathfrak{N}$, the class of nilpotent groups, is that it is $\boldsymbol{N}_0$-closed. More precisely there is the well-known result

Theorem 2.18 (Fitting [1], p. 100). Let $H \lhd G$ and $K \lhd G$ and assume that H and K are nilpotent of classes c and d respectively. Then HK is nilpotent of class at most $c + d$

Proof. If A, B and C are normal subgroups of a group, it follows easily from equation (2) on p. 42 that

$$[AB, C] = [A, C]\,[B, C].$$

This enables us to prove by induction on t that

$$\gamma_{t+1}(HK) = \prod\, [L_1, \ldots, L_{t+1}]$$

where $L_i = H$ or K and all possible commutator subgroups of this type are included in the product. Let $t = c + d$: then there are at least $c + 1$ H's or $d + 1$ K's among $L_1, \ldots, L_{t+1}$ and, since $H \lhd G$ and $K \lhd G$, the subgroup $[L_1, \ldots, L_{t+1}]$ is contained in either $\gamma_{c+1}(H)$ or $\gamma_{d+1}(K)$, both of which are trivial. Hence $\gamma_{c+d+1}(HK) = 1$. $\square$

Notice that while the $\boldsymbol{N}_0$-closure of the class $\mathfrak{S}$ arises from its $\boldsymbol{P}$-closure, the class $\mathfrak{N}$ is not $\boldsymbol{P}$-closed.

Several generalizations of nilpotency suggest themselves. A *Z-group* is a group with a central series. A group with an ascending central series is called *hypercentral* (or a *ZA-group*) and a group with a descending central series is called a *hypocentral* (or *ZD*)-*group*. Also there is the class of *locally nilpotent groups*, which will be of recurring interest to us.

It is well-known that finite nilpotent groups can be characterized by many properties which for infinite groups lead to generalizations of nilpotency. One example is the property that every subgroup is subnormal: if G is any nilpotent group and H is a subgroup of G, then

$$[\zeta_{i+1}(G), H] \leqq \zeta_i(G);$$

hence $H\zeta_i(G) \lhd H\zeta_{i+1}(G)$ and $H \lhd^c G$ where c is the nilpotent class of G. The same argument shows that *every subgroup of a hypercentral group is ascendant.* A survey of the classes of generalized nilpotent groups is given in Chapter 6.

Hypercentral Groups

We will conclude this section with some properties of hypercentral groups from which it will be seen that there is a certain analogy with hyperabelian groups. Clearly, a group is hypercentral if and only if it coincides with its hypercentre. Also, *a group is hypercentral if and only if each non-trivial homomorphic image has non-trivial centre* (Lemma 1.22).

Corresponding to Baer's criterion for hyperabelian groups (Theorem 2.15) there is

Theorem 2.19 (Černikov [20], Theorem 1). A group G is hypercentral if and only if given two sequences $x_1, x_2, \ldots$ and $y_1, y_2, \ldots$ of elements of G such that

$$x_{i+1} = [x_i, y_i],$$

there is an integer $m > 0$ such that $x_m = 1$.

The proof, which is similar to that of Theorem 2.15, is left to the reader. A corollary is that *a group is hypercentral if and only if its countable subgroups are hypercentral* (Černikov [20], Theorem 2).

The next result is frequently used.

Lemma 2.19.1. Let A and N be normal subgroups of a hypercentral group G and assume that $A \leq N$. Then A is a maximal G-admissible abelian subgroup of N if and only if $A = C_N(A)$.

Proof. If $A = C_N(A)$, then A is abelian and cannot lie in a larger abelian subgroup. Conversely, let A be a maximal G-admissible abelian subgroup of N and assume that $A < C = C_N(A)$. Since $C \lhd G$, Lemma 2.16 implies that C/A contains a non-trivial element zA belonging to the centre of G/A. But $\langle z, A\rangle$ is abelian and is contained in N, while $\langle z, A\rangle \lhd G$ since $[z, G] \leq A$; thus $z \in A$, which is contrary to the choice of z. □

We conclude with two further properties of hypercentral groups.

a) Hypercentral Groups are Locally Nilpotent (Mal'cev [4], p. 212)

We begin with the well-known remark: *finitely generated nilpotent groups are polycyclic.* This is a consequence of Theorem 2.26 and can also be proved as follows. By repeated use of Lemma 2.12 we see that if $G = \langle X\rangle$, then $\gamma_i(G)$ can be generated modulo $\gamma_{i+1}(G)$ by all the commutators $[x_1, \ldots, x_i]$ with $x_j \in X$; hence if G is a finitely generated nilpotent group, each lower central factor is finitely generated and G is polycyclic.

Let G be a hypercentral group and set $Z_\alpha = \zeta_\alpha(G)$. If X is an arbitrary finitely generated subgroup of G, let α be the least ordinal such that $X/X \cap Z_\alpha$ is nilpotent; there are ordinals with this property since $G = Z_{\alpha_1}$ for some α_1. If α is not a limit ordinal, $X \cap Z_\alpha/X \cap Z_{\alpha-1}$ lies in the centre of $X/X \cap Z_{\alpha-1}$ and therefore $X/X \cap Z_{\alpha-1}$ is nilpotent. Hence α must be a limit ordinal; if $\alpha = 0$, then X is nilpotent and we are finished, so let $\alpha > 0$. Now $X/X \cap Z_\alpha$ is polycyclic, and therefore finitely presented, by the Corollary to Lemma 1.43. The same lemma implies that $X \cap Z_\alpha$ is finitely generated as an X-operator group. But this shows that $X \cap Z_\alpha = X \cap Z_\beta$ for some $\beta < \alpha$, which again contradicts the minimality of α. ▯

For another proof see Kuroš [9], § 63.

b) The Hypercentral Groups Form an $\mathbf{N}_0$*-Closed Class* (P. Hall [11], Lemma 1)

Let H and K be two normal hypercentral subroups of a group and let $J = HK$. Evidently it is enough to show that if $J \neq 1$, then $\zeta(J) \neq 1$. We can assume that $H \neq 1$, so that $\zeta(H) \neq 1$. If $H \cap K = 1$, we have $[H, K] = 1$ and $\zeta(H) \leqq \zeta(J)$. If $H \cap K \neq 1$, then, since $H \cap K \lhd H$, it follows that $1 \neq (H \cap K) \cap \zeta(H) = \zeta(H) \cap K$ by Lemma 2.16. Since $1 \neq \zeta(H) \cap K \lhd K$, we conclude for the same reason that

$$1 \neq (\zeta(H) \cap K) \cap \zeta(K) = \zeta(H) \cap \zeta(K).$$

Clearly $\zeta(H) \cap \zeta(K) \leqq \zeta(J)$. ▯

2.2 Properties of the Upper and Lower Central Series

Lemma 2.21 (P. Hall [1], pp. 53—55). Let G be a group and let m and n be non-negative integers.

(i) $\zeta_m(G/\zeta_n(G)) = \zeta_{m+n}(G)/\zeta_n(G)$.

(ii) $[\gamma_m(G), \gamma_n(G)] \leqq \gamma_{m+n}(G)$, $(m, n > 0)$.

(iii) $[\gamma_m(G), \zeta_n(G)] \leqq \zeta_{n-m}(G)$, $(0 < m \leqq n)$.

(iv) $\gamma_m(\gamma_n(G)) \leqq \gamma_{mn}(G)$, $(m, n > 0)$.

These results may be proved by induction on m: in the case of the last three the Three Subgroup Lemma is required. Part (ii) has a well-known consequence.

Corollary. If G is any group and m is a non-negative integer, $G^{(m)} \leqq \gamma_{2^m}(G)$. If G is nilpotent of positive class c, the derived length of G is at most $[\log_2 c] + 1$.

Lemma 2.22 (D. H. McLain [5]). Let $X \leqq G$, let $N \lhd G$ and suppose that $G = XN'$. Then if i is a positive integer, $G = X\gamma_i(N)$.

Proof. Suppose that $G = X\gamma_i(N)$ where $i \geqq 2$: then

$$N = N \cap (X\gamma_i(N)) = (N \cap X)\,\gamma_i(N).$$

This implies that for any positive integer j

$$\gamma_j(N) \leqq \gamma_j(N \cap X)\,\gamma_{i+j-1}(N). \tag{4}$$

For if we assume that this is true for j and commute both sides with $N = (N \cap X)\,\gamma_i(N)$, we obtain $\gamma_{j+1}(N) \leqq \gamma_{j+1}(N \cap X)\,\gamma_{i+j}(N)$, by using equations (1) and (2) on p. 42 and the inclusion

$$[\gamma_j(N \cap X), \gamma_i(N)] \leqq \gamma_{i+j}(N)$$

(Lemma 2.21). Thus (4) is true by induction on j. Set $j = i$ in (4) and substitute for $\gamma_i(N)$ in $G = X\gamma_i(N)$: we obtain

$$G = X\gamma_{2i-1}(G) = X\gamma_{i+1}(G)$$

since $i \geqq 2$. ∎

The best known case of this lemma is when $N = G$ and the assertion is that a subgroup which generates G modulo G' also generates G modulo any term of the lower central series of G with finite ordinal type.

Upper Central Factors

It frequently happens that we have information about the centre or the derived factor group of a group and we wish to transfer this information to other factors of the upper or lower central series respectively — or to those of finite ordinal type at least. Consider first the case of the upper central series.

Theorem 2.23 (Dixmier [1]). If G is a group whose centre has finite exponent e, then $\zeta_{i+1}(G)/\zeta_i(G)$ has exponent dividing e for each positive integer i.

Proof. It is enough to show that $\zeta_2(G)/\zeta_1(G)$ has exponent dividing e. Let $x \in \zeta_2(G)$ and let $g \in G$; then $[x, g] \in \zeta_1(G)$, so by the elementary formula (2) (p. 42),

$$1 = [x, g]^e = [x^e, g].$$

Hence $x^e \in \zeta_1(G)$ as required. ∎

Corollary. If G is a nilpotent group of class c whose centre has finite exponent e, then G has exponent dividing e^c.

Theorem 2.24. A finitely generated nilpotent group G with periodic centre is finite.

Proof. Let $g_1, \dots, g_n$ generate G and let $x \in \zeta_2(G)$. Each $[x, g_i]$ belongs to the centre of G and thus has finite order. Let l be the least common multiple of the orders of the $[x, g_i]$. Then $1 = [x, g_i]^l = [x^l, g_i]$, so $x^l \in \zeta_1(G)$ and $\zeta_2(G)/\zeta_1(G)$ is periodic. This argument shows that G is periodic. Hence G is locally finite, by Theorem 1.45, and therefore G is finite. $\square$

On the other hand, *there exist non-periodic nilpotent groups with periodic centre.* For example, let A be an additive abelian group which is periodic but has infinite exponent. Let D be the direct sum $A \oplus A$. The mapping α defined by

$$(a_1, a_2)\alpha = (a_1 + a_2, a_2), \quad (a_i \in A),$$

is an automorphism of D; furthermore, α has infinite order since $(a_1, a_2)\alpha^m = (a_1 + ma_2, a_2)$ and A has infinite exponent. Let G be the holomorph of D by $\langle\alpha\rangle$. Then $\zeta(G)$ and G' coincide with the set of all $(a_1, 0)$, which is a subgroup isomorphic with A. Thus G is nilpotent of class 2, but is not periodic.

Although periodicity is not a transferable property for the upper central series, McLain [4]—and later Kontorovič [8]—has shown that the property of being torsion-free is inherited from the centre by *every* upper central factor. Special cases of this were found previously by Mal'cev [4] and Černikov [19].

In fact McLain proved a more general result. If π is a set of primes, a group is said to be *π-free* if it contains no non-trivial elements whose order is a *π-number* (i.e. a product of primes in π).

Theorem 2.25. Let G be a group with π-free centre. Then each upper central factor, and therefore the hypercentre of G, is π-free.

Proof. Suppose that the theorem is false and let α be the first ordinal such that $\zeta_{\alpha+1}(G)/\zeta_\alpha(G)$ is not π-free. Then $\alpha > 0$ and there is an element $x \in \zeta_{\alpha+1}(G)\backslash\zeta_\alpha(G)$ such that $x^m \in \zeta_\alpha(G)$ for some positive π-number m. Assume that α is not a limit ordinal. For any $g \in G$, we have $[g, x] \in \zeta_\alpha(G)$ and $[g, x]^m \equiv [g, x^m] \equiv 1 \bmod \zeta_{\alpha-1}(G)$. But $\zeta_\alpha(G)/\zeta_{\alpha-1}(G)$ is π-free, so $[g, x] \in \zeta_{\alpha-1}(G)$ and $x \in \zeta_\alpha(G)$, which is a contradiction. Hence α must be a limit ordinal and $x^m \in \zeta_{\beta+1}(G)\backslash\zeta_\beta(G)$ for some $\beta < \alpha$. Let $g \in G$: then $[g, x] \in \zeta_{\gamma+1}(G)\backslash\zeta_\gamma(G)$ where $\gamma = \gamma(g) < \alpha$. Since $x \notin \zeta_\alpha(G)$, there is g such that $\gamma(g) \geqq \beta$. With this g we have $[g, x]^m \equiv [g, x^m] \bmod \zeta_\gamma(G)$ and also $[g, x^m] \in \zeta_\beta(G) \leqq \zeta_\gamma(G)$. Hence $[g, x]^m \in \zeta_\gamma(G)$. But $[g, x] \in \zeta_{\gamma+1}(G)$ and $\zeta_{\gamma+1}(G)/\zeta_\gamma(G)$ is π-free, so $[g, x] \in \zeta_\gamma(G)$, which contradicts the definition of γ. $\square$

Lower Central Factors

We come now to the question: which properties of the derived factor group of a group are inherited by other lower central factors of finite type? The discussion which follows is taken from a paper of Robinson [10].

Let G be a group and let Σ be a group of operators on G. For brevity let $G = G_1, G_2, \ldots$ denote the terms of the lower central series of G with finite type. Then

$$G_{i+1} = [G_i, G].$$

Obviously each G_i is Σ-admissible, so G_i/G_{i+1} can be regarded as a Σ-group in a natural way. Of course G_i/G_{i+1} is an abelian group, so we can make it into a module over $Z\Sigma$, the group ring of Σ over the ring of integers Z, according to the rule

$$(aG_{i+1})^r = \prod_{j=1}^{k} (a^{\sigma_j})^{n_j} G_{i+1}$$

where

$$r = \sum_{j=1}^{k} n_j\sigma_j \in Z\Sigma, \quad (n_j \in Z, \sigma_j \in \Sigma).$$

(The fact that G_i/G_{i+1} is written multiplicatively is, of course, irrelelant).

The next step is to consider how these modules G_i/G_{i+1} are related. Choose a fixed integer i and for any $a \in G_i$ and $g \in G$ write

$$\bar{a} = aG_{i+1} \text{ and } g^* = gG'.$$

Let F be the free multiplicative abelian group on the set of all pairs $(\bar{a}, g^*)$, $a \in G_i$, $g \in G$. Then the mapping

$$(\bar{a}, g^*) \to [a, g]\, G_{i+2} \tag{5}$$

is well-defined: this is because $[G_i, G'] \leq G_{i+2}$ by Lemma 2.21. The mapping (5) extends to a homomorphism of F onto G_{i+1}/G_{i+2} and, by (1) and (2), on p. 42, elements in F which have the form

$$(\bar{a}_2, g^*)^{-1}(\bar{a}_1, g^*)^{-1}(\bar{a}_1\bar{a}_2, g^*)$$

or

$$(\bar{a}, g_1^*g_2^*)\,(\bar{a}, g_1^*)^{-1}\,(\bar{a}, g_2^*)^{-1}$$

lie in the kernel of the homomorphism. Hence, by definition of the tensor product over Z, the mapping

$$\bar{a} \otimes g^* \to [a, g]\, G_{i+2} \tag{6}$$

defines a homomorphism θ of

$$(G_i/G_{i+1}) \otimes (G/G')$$

onto G_{i+1}/G_{i+2}. There is a natural way of making the tensor product over Z of two $Z\Sigma$-modules A and B into a $Z\Sigma$-module: we simply define

$$(a \otimes b)\,\sigma = (a\sigma) \otimes (b\sigma),\ a \in A,\ b \in B,\ \sigma \in \Sigma.$$

A routine argument shows that this action of σ is well-defined and that θ is a $Z\Sigma$-homomorphism. Consequently *the mapping (6) defines a $Z\Sigma$-homomorphism of $(G_i/G_{i+1}) \otimes (G/G')$ onto G_{i+1}/G_{i+2}.* (This result can be sharpened by using a skew tensor product to reflect the commutator relation $[x, y]\,[y, x] = 1$).

These observations permit us to draw a number of conclusions about the influence of G/G' on the G_i/G_{i+1}. The following result is obvious if we take Σ to be trivial in the above discussion.

Theorem 2.26 (Robinson [10]). Let $\mathfrak{X}$ be a class of groups which contains every homomorphic image of the tensor product of two abelian $\mathfrak{X}$-groups. If G is a group such that G/G' belongs to $\mathfrak{X}$, then $\gamma_i(G)/\gamma_{i+1}(G)$ belongs to $\mathfrak{X}$ for each finite i.

Corollary. If in addition G is nilpotent and $\mathfrak{X}$ is **P**-closed, then $G/G' \in \mathfrak{X}$ implies that $G \in \mathfrak{X}$.

We will mention some of the possibilities for $\mathfrak{X}$ in this corollary: finite π-groups, periodic groups, groups of finite rank, groups satisfying Max: except in the third case these results are due to Baer ([11], p. 361—364). Bearing in mind that finitely generated abelian groups satisfy Max, we see that the third result of Baer asserts that *a nilpotent group G for which G/G' is finitely generated satisfies Max and so is polycyclic.* By taking for $\mathfrak{X}$ the class of finite p-groups one obtains the following: a nilpotent group that is generated by elements each of which has order a power of a prime p is a p-group. This has the important consequence that *a locally nilpotent group G has a unique Sylow p-subgroup for each prime p and the elements of finite order in G form a subgroup which is the direct product of the Sylow subgroups.*

Another possibility for $\mathfrak{X}$ is the class of groups that satisfy Min. The tensor product of an abelian group of finite exponent with any other abelian group still has finite exponent; therefore the same method of proof establishes a result of Blackburn ([2], Theorem 1.5) and Dixmier [1]: *if the αth lower central factor of a group has finite exponent e, the $(\alpha + n)$th lower central factor has exponent dividing e for all finite n.* A related result of Lam [1] is equally easy to prove.

None of the applications so far has exploited the full module structure of the tensor product. We will rectify this situation and prove a theorem of P. Hall which is usually proved by commutator calculations.

Let G be a group and let $N \lhd G$. Then N admits G as a group of operators and N_i/N_{i+1}, the ith lower central factor of N, is a ZG-module. Our general considerations show that N_{i+1}/N_{i+2} is a ZG-homomorphic image of the ZG-module $(N_i/N_{i+1}) \otimes (N/N')$.

Theorem 2.27 (P. Hall [7], Theorem 7). Let N be a normal subgroup of a group G and suppose that N and G/N' are nilpotent with classes c and d respectively. Then G is nilpotent of class at most $\binom{c+1}{2}d - \binom{c}{2}$

Proof. Since G/N' is nilpotent, there is a G-central series of length at most d from N' to N. The factors of this series are trivial ZG-modules, so we can say that N/N' is *polytrivial* of length at most d as a ZG-module. Obviously a homomorphic image of a polytrivial ZG-module is also polytrivial and of length not greater than that of the original module. What has to be established is that polytriviality is a property which passes to tensor products over Z.

Let A and B be polytrivial ZG-modules of length $m > 0$ and $n > 0$ respectively and let $T = A \otimes B$: here G can be any group. Then there are series of submodules

$$0 = A_0 < A_1 < \cdots < A_m = A \text{ and } 0 = B_0 < B_1 < \cdots < B_n = B$$

such that A_{i+1}/A_i and B_{j+1}/B_j are trivial ZG-modules. Let

$$T_k = \langle a \otimes b : a \in A_i, b \in B_j, i + j \leqq k \rangle.$$

If $g \in G$, then $(a \otimes b)\, g = (ag) \otimes (bg)$, so T_k is a submodule of T and

$$0 = T_1 \leqq T_2 \leqq \cdots \leqq T_{m+n} = T$$

is a series of submodules in T with length at most $m + n - 1$. The mapping

$$(a + A_i, b + B_j) \to (a \otimes b) + T_k,$$

where $a \in A_{i+1}, b \in B_{j+1}$ and $i + j = k - 1$, determines a homomorphism of the trivial module $(A_{i+1}/A_i) \otimes (B_{j+1}/B_j)$ into T_{k+1}/T_k. The latter is the sum of all the homomorphic images for $i + j = k - 1$, so

$$T_{k+1}/T_k$$

is trivial and T is polytrivial with length at most

$$m + n - 1.$$

Consider now the situation of the theorem and suppose that N_i/N_{i+1}, the ith lower central factor of N, is polytrivial of length l as a ZG-module. Then N_{i+1}/N_{i+2} is polytrivial of length at most $l + d - 1$. Therefore, by induction on i, the module N_i/N_{i+1} is polytrivial of length at most $d + (i - 1)(d - 1) = i(d - 1) + 1$ and there is a G-central series in N' of length at most

$$\sum_{i=2}^{c} (i(d-1)+1).$$

It follows that G is nilpotent with class not exceeding

$$d + \sum_{i=2}^{c} (i(d-1)+1) = \binom{c+1}{2} d - \binom{c}{2}. \quad \square$$

However, this bound for the class of G is not best possible: indeed the best possible bound was shown by A. G. R. Stewart [1] to be

$$cd + (c-1)(d-1).$$

This is obtainable by slightly more elaborate commutator calculations, including use of equation (8) on p. 117.

Other theorems of "Hall's type" can be proved by similar methods: for example, we could replace the property of being a trivial ZG-module by that of having a cyclic underlying additive abelian group, or we could allow well-ordered ascending series of submodules instead of merely series of finite length. The following result has been obtained (Robinson [10]).

If N is a normal nilpotent subgroup of a group G and if G/N' has a property $\mathcal{P}$, then G has the property $\mathcal{P}$ where $\mathcal{P}$ is supersoluble, hypercentral, hypercyclic, locally nilpotent, locally supersoluble or locally polycyclic. In the locally nilpotent case this result was first obtained by Plotkin in the paper [18] which also contains a proof of Hall's original theorem. Subsequently in [1] Betten gave another proof of the hypercentral and hypercyclic cases.

2.3 The Hirsch-Plotkin-Baer Theorem and Related Results

The $\boldsymbol{N}_0$-closure of the class of nilpotent groups (Lemma 2.18) was established by Fitting in 1938. The deeper result that the class of locally nilpotent groups is $\boldsymbol{N}_0$-closed was proved independently by Hirsch [7] and Plotkin [6] some fifteen years later and has since turned out to be of fundamental importance. Here we will establish a theorem of a more general type, which also incorporates a result of Baer. This is

Theorem 2.31. Let $\mathfrak{X}$ be a class of groups which is $\boldsymbol{N}_0$ and $\boldsymbol{S}$-closed and such that every $\mathfrak{X}$-group is finitely generated. Then the class of locally $\mathfrak{X}$-groups is $\acute{\boldsymbol{N}}$-closed.

Proof. Since $\acute{\boldsymbol{N}} \leqq \langle \boldsymbol{L}, \boldsymbol{N}_0 \rangle$ by equation (10) on p. 20, it is enough to prove that $\boldsymbol{L}\mathfrak{X}$ is $\boldsymbol{N}_0$-closed.

Let H and K be normal locally $\mathfrak{X}$-subgroups of some group. In order to prove that HK is locally-$\mathfrak{X}$ it need only be shown that if X and Y are finitely generated subgroups of H and K respectively, then $J = \langle X, Y \rangle \in \mathfrak{X}$. Note that $X \in \mathfrak{X}$ and $Y \in \mathfrak{X}$ since $\mathfrak{X} = \boldsymbol{S}\mathfrak{X}$. Let $X = \langle x_1, \ldots, x_m \rangle$ and $Y = \langle y_1, \ldots, y_n \rangle$. Then the subgroup

$$Z = \langle [x_i, y_j]: \quad i = 1, \ldots, m, \quad j = 1, \ldots, n \rangle$$

is contained in $H \cap K$ since H and K are normal subgroups. Thus $\langle X, Z \rangle$ is a finitely generated subgroup of H, so it belongs to $\mathfrak{X}$. Hence $Z^X \in \mathfrak{X}$ and consequently Z^X is finitely generated. Therefore $\langle Y, Z^X \rangle$ is a finitely generated subgroup of K and belongs to $\mathfrak{X}$. But $[X, Y] = Z^{XY}$, by the corollary to Lemma 2.12, so

$$Y^X = \langle Y, [X, Y] \rangle = \langle Y, Z^{XY} \rangle = \langle Y, Z^X \rangle.$$

Hence $Y^X \in \mathfrak{X}$ and in a similar way $X^Y \in \mathfrak{X}$. But $X^Y \lhd J$ and $Y^X \lhd J$, so

$$J = X^Y Y^X \in \boldsymbol{N}_0 \mathfrak{X} = \mathfrak{X}. \quad \square$$

Since finitely generated nilpotent groups satisfy Max and constitute an $\boldsymbol{S}$ and $\boldsymbol{N}_0$-closed class, $\mathfrak{G} \cap \mathfrak{N}$ is a possibility for $\mathfrak{X}$ here: in this case the conclusion is

$$\acute{\boldsymbol{N}}(\boldsymbol{L}\mathfrak{N}) = \boldsymbol{L}\mathfrak{N}, \tag{7}$$

which implies the original theorem of Hirsch and Plotkin. $\mathfrak{X}$ could also be the class of groups which satisfy Max (by Lemma 1.48), in which case we find that the class of groups satisfying Max locally is $\acute{\boldsymbol{N}}$ and therefore $\boldsymbol{N}_0$-closed (Baer [27], § 2, Satz 1).

The above proof is basically that of P. Hall ([6], Theorem 2.8): for a proof using basic commutators see Gruenberg [3]. A slightly different form of Theorem 2.31 appears in the survey paper of Plotkin [12] (2.4.2) and also in Baer [27] (§ 4, Satz 1).

Radical Groups

If G is any group, the product of all the normal locally nilpotent subgroups of G is the radical

$$\varrho_{\boldsymbol{L}\mathfrak{N}}(G). \tag{8}$$

Theorem 2.31 implies at once that this subgroup is locally nilpotent, and hence is the unique largest maximal normal locally nilpotent subgroup of G. This important subgroup is called the *Hirsch-Plotkin radical of* G. By Theorem 2.31 the Hirsch-Plotkin radical contains all the *ascendant* locally nilpotent subgroups.

However the class of locally nilpotent groups is not a radical class, since it is not $\boldsymbol{P}$-closed. The identity of the radical class generated by the class of locally nilpotent groups is easily settled:

$$\operatorname{Rad}(\boldsymbol{L}\mathfrak{N}) = \acute{\boldsymbol{P}}_n\boldsymbol{L}\mathfrak{N} = \acute{\boldsymbol{P}}_{sn}\boldsymbol{L}\mathfrak{N} = \acute{\boldsymbol{P}}\boldsymbol{L}\mathfrak{N}.$$

This is because $\operatorname{Rad}(\boldsymbol{L}\mathfrak{N}) = \acute{\boldsymbol{P}}_{sn}\boldsymbol{L}\mathfrak{N}$, by Theorem 1.33, and because an ascendant locally nilpotent subgroup always lies in a normal locally nilpotent subgroup, for example the Hirsch-Plotkin radical. Thus $\operatorname{Rad}(\boldsymbol{L}\mathfrak{N})$ is the class of hyper-(locally nilpotent) groups and it is also the class of groups G in which the upper locally nilpotent series or series of iterated Hirsch-Plotkin radicals reaches G.

A group which has one of these equivalent properties is called a *radical group*, a terminology due to Plotkin. Notice that among the radical groups are all locally nilpotent groups and soluble groups. *Periodic radical groups are locally finite*: this is proved by noting that finitely generated periodic nilpotent groups are finite—by Theorem 2.24 for example—and that the class of locally finite groups is $\acute{\boldsymbol{P}}$-closed, by the Corollary to Theorem 1.45. Plotkin studies radical groups in detail in his paper [7] and some of his results appear in later chapters. For the present we merely note

Lemma 2.32 (Plotkin [7], Theorem 7). Let G be a radical group and let H be its Hirsch-Plotkin radical. Then $C_G(H) \leqq H$.

This is analogous to Lemma 2.17 and may be proved in a similar way. One consequence is that a *radical group whose Hirsch-Plotkin radical is finite is a finite soluble group.*

Theorems of Baer, Gruenberg and Wielandt

A class of groups $\mathfrak{X}$ is said to be an *ascendant coalition class* if in any group a pair of ascendant $\mathfrak{X}$-subgroups generates an ascendant $\mathfrak{X}$-subgroup. The definition of a *subnormal coalition class* is obtained if we replace the word "ascendant" by "subnormal".

Theorem 2.33. Let $\mathfrak{X}$ be a class of groups which is $\boldsymbol{N}_0$ and $\boldsymbol{S}$-closed and such that every $\mathfrak{X}$-group is finitely generated. Then $\mathfrak{X}$ is both an ascendant coalition class and a subnormal coalation class.

Special choices of $\mathfrak{X}$ here lead to various known theorems. Thus the class of finite π-groups and the class of polycyclic groups are subnormal coalition classes (Wielandt [2], p. 218): the class of finitely generated nilpotent groups is an ascendant coalition class (Gruenberg [3], Theorem 2) and a subnormal coalition class (Baer [20], § 3, Satz 3). Other examples of coalition classes are to be found in papers of Roseblade [3], [5]; Robinson [4], [14] and Roseblade and Stonehewer [1].

Proof of Theorem 2.33. Let H and K be ascendant $\mathfrak{X}$-subgroups of a group G and let $J = \langle H, K \rangle$. Theorem 2.31 shows that J is a locally $\mathfrak{X}$-group: on the other hand, H and K are finitely generated, being $\mathfrak{X}$-groups, so J is also finitely generated and hence $J \in \boldsymbol{S}\mathfrak{X} = \mathfrak{X}$. We still have to prove that J *asc* G.

Let

$$H = V_0 \lhd V_1 \lhd \cdots V_\alpha = G$$

be an ascending series from H to G with $\alpha > 0$. Now $H^J \in \boldsymbol{S}\mathfrak{X} = \mathfrak{X}$, so H^J is finitely generated and hence may be generated by finitely many conjugates of H in J. If α is a limit ordinal, it follows that $H^J \leqq V_\beta$ for some $\beta < \alpha$, and, since $H^x \lhd^\beta V_\beta$ for $x \in J$, transfinite induction on α permits us to conclude that H^J *asc* V_β and therefore that H^J *asc* G. If α is not a limit ordinal, $H \leqq V_{\alpha-1} \lhd G$, so $H^J \leqq V_{\alpha-1}$: just as before, H^J *asc* $V_{\alpha-1}$ and H^J *asc* G, by transfinite induction. Consequently H^J *asc* G and $H^J \in \mathfrak{X}$ in either case, so we may suppose that

$$H \lhd J.$$

Now define

$$\overline{V}_\beta = H^{V_\beta}$$

for $\beta \leqq \alpha$. Then $H = \overline{V}_0 = \overline{V}_1 \lhd \overline{V}_2 \lhd \cdots \overline{V}_\alpha = H^G$ is an ascending series from H to H^G. Also $H^G \in \boldsymbol{L}\mathfrak{X}$ by Theorem 2.31, so $\overline{V}_\beta \in \boldsymbol{L}\mathfrak{X}$ for each $\beta \leqq \alpha$. To obtain a K-invariant series, set

$$W_\beta = \bigcap_{x \in K} (\overline{V}_\beta)^x = \mathrm{Core}_K\, \overline{V}_\beta.$$

Then one easily verifies that $W_\beta \lhd W_{\beta+1}$, $H = W_0 = W_1$ and $H^G = W_\alpha$. However, in order to conclude that the W_β's form an ascending series from H to H^G we must check that

$$W_\lambda = \bigcup_{\beta<\lambda} W_\beta$$

for every limit ordinal $\lambda \leqq \alpha$. One inclusion, $W_\lambda \geqq \bigcup_{\beta<\lambda} W_\beta$, is obvious. Let $x \in W_\lambda$. Since $\overline{V}_\lambda$ and K are ascendant $\boldsymbol{L}\mathfrak{X}$-groups, $\langle \overline{V}_\lambda, K \rangle \in \boldsymbol{L}\mathfrak{X}$, the class $\boldsymbol{L}\mathfrak{X}$ being $\acute{\boldsymbol{N}}$-closed. $x \in W_\lambda \leqq \overline{V}_\lambda$, so $\langle x, K \rangle$ is a finitely generated subgroup of $\langle \overline{V}_\lambda, K \rangle$. Hence $\langle x, K \rangle$, and also x^K, is an $\mathfrak{X}$-group: consequently, x^K is finitely generated and is contained in some $\overline{V}_\beta$, where $\beta < \lambda$, since $x^K \leqq W_\lambda^K = W_\lambda \leqq \overline{V}_\lambda$. Therefore $x^K \leqq V_\beta^y$ for all $y \in K$, so that $x^K \leqq W_\beta$ and $x \in W_\beta$, as required.

The remainder of the proof is straightforward. For each $\beta \leqq \alpha$ we have $W_\beta \lhd KW_{\beta+1}$ and of course K *asc* $KW_{\beta+1}$. Ascendance is preserved under homomorphisms, so KW_β *asc* $KW_{\beta+1}$ and $J = KW_0$ *asc* $KW_\alpha = KH^G$. But K *asc* G and $H^G \lhd G$ imply that KH^G *asc* G, so J *asc* G.

The proof that $\mathfrak{X}$ is a subnormal coalition class is similar but easier since no discussion of limit ordinals is required. ▯

Baer Groups and Gruenberg Groups

Since the class $\boldsymbol{L}\mathfrak{N}$ is $\acute{\boldsymbol{N}}$-closed, $\acute{\boldsymbol{N}}\mathfrak{A} \leqq \boldsymbol{L}\mathfrak{N}$. Also in a nilpotent group every subgroup is subnormal. Hence, by Corollary 1 to Lemma 1.31, the inclusion $\mathfrak{N} \leqq \boldsymbol{N}\mathfrak{A}$ is valid. Thus

$$\mathfrak{N} \leqq \boldsymbol{N}\mathfrak{A} \leqq \acute{\boldsymbol{N}}\mathfrak{A} \leqq \boldsymbol{L}\mathfrak{N} \tag{9}$$

and we have found two new classes of generalized nilpotent groups, both subclasses of the class of locally nilpotent groups. The class

$$\boldsymbol{N}\mathfrak{A}$$

consists of groups generated by their subnormal abelian subgroups and such groups are called *Baer groups* (or *Baer nil-groups*). The class

$$\acute{\boldsymbol{N}}\mathfrak{A}$$

consists of groups generated by their ascendant abelian subgroups: we will call these *Gruenberg groups*.

Lemma 2.34. The following properties of a group G are equivalent.

(i) Every finitely generated subgroup of G is subnormal (ascendant).
(ii) Every cyclic subgroup of G is subnormal (ascendant).
(iii) G is a Baer (Gruenberg) group.

Proof. It is obvious that (i) implies (ii) and (ii) implies (iii). Let G be a Baer (Gruenberg) group: then a finitely generated subgroup X of G lies inside the join of a finite number of finitely generated subnormal (ascendant) abelian subgroups, i.e. inside a subnormal (ascendant) finitely generated nilpotent subgroup of G, by Theorem 2.33. Hence X is subnormal (ascendant) in G. ▯

In any group there is a *Baer radical* which is a Baer group and contains all subnormal Baer subgroups and a *Gruenberg radical*, which is a Gruenberg group and contains all ascendant Gruenberg subgroups. These radicals are intermediate between the Fitting subgroup and the Hirsch-Plotkin radical.

Finally we observe the following connections between these classes of generalized nilpotent groups and the classes of subsoluble and SN^*-groups.

Lemma 2.35. (i) A group G is subsoluble if and only if the upper Baer series terminates at G.

(ii) A group G is SN^* if and only if the upper Gruenberg series terminates at G.

Proof. A group is subsoluble if and only if non-trivial homomorphic images contain non-trivial subnormal abelian subgroups (Lemma 1.23). Hence every Baer group is subsoluble and if the upper Baer series reaches G, then G is subsoluble. Since the Baer radical contains all subnormal abelian subgroups, a group which is subsoluble must coincide with a term of its upper Baer series. Thus (i) is established: the proof of (ii) is similar. ▯

An Analogue of the Hirsch-Plotkin-Baer Theorem

If $\mathfrak{X}$ is a class of groups, we recall that

$$\boldsymbol{M}\mathfrak{X}$$

is the class of locally normal and $\mathfrak{X}$-groups. The operation $\boldsymbol{M}$ generates a closure operation $\overline{\boldsymbol{M}}$ where, as one readily shows,

$$\overline{\boldsymbol{M}}\mathfrak{X} = \bigcup_{\alpha} \boldsymbol{M}^{\alpha}\mathfrak{X} \tag{10}$$

—see Section 1.1, p. 13. It is clear that

$$\boldsymbol{M} \leqq \overline{\boldsymbol{M}} \leqq \boldsymbol{L}.$$

Suppose that $\mathfrak{X}$ is an $\boldsymbol{S}$-closed class; then, by (10) and the definition of $\boldsymbol{M}$, a group belongs to $\overline{\boldsymbol{M}}\mathfrak{X}$ if and only if, given a descending series

$$G = G_1 \geqq G_2 \geqq \cdots$$

in which

$$G_{i+1} = X_i^{G_i} \quad (i = 1, 2, \ldots)$$

for some finite subset X_i of G_i, there is a positive integer n such that $G_n \in \mathfrak{X}$.

The following result, comparable with the Hirsch-Plotkin-Baer Theorem, has been established by Rips [1] under the additional hypothesis that $\mathfrak{X}$ is $\boldsymbol{H}$-closed.

Theorem 2.36. Let $\mathfrak{X}$ be a class of groups which is $\boldsymbol{N}_0$ and $\boldsymbol{S}$-closed and such that every $\mathfrak{X}$-group is finitely generated. Then the class $\overline{\boldsymbol{M}}\mathfrak{X}$ is $\boldsymbol{N}$-closed.

Proof. If a group G is a product of normal $\overline{\boldsymbol{M}}\mathfrak{X}$-groups, each finite subset of G lies in the product of a finite set of these normal subgroups. Thus we need only prove that $\overline{\boldsymbol{M}}\mathfrak{X}$- is $\boldsymbol{N}_0$-closed.

Let H and K be two normal $\bar{M}\mathfrak{X}$-subgroups of a group and let $J = HK$. We have to prove that $J \in \bar{M}\mathfrak{X}$. We know by (10) that H and K belong to $M^{\alpha}\mathfrak{X}$ for some ordinal α. If $\alpha = 0$ then H and K belong to $\mathfrak{X}$ and $J \in N_0\mathfrak{X} = \mathfrak{X}$. Thus we may suppose that $\alpha > 0$. Let us make the inductive hypothesis that the product of any pair of normal $M^{\beta}\mathfrak{X}$-subgroups belongs to $\bar{M}\mathfrak{X}$ for each $\beta < \alpha$. If J is not in $\bar{M}\mathfrak{X}$, there must exist a descending series

$$J = J_1 \geqq J_2 \geqq \cdots$$

in which

$$J_{i+1} = X_i^{J_i} \quad (i = 1, 2, \ldots) \tag{11}$$

for some finite subset X_i of J_i and in which no J_i belongs to $\mathfrak{X}$. It is clear that in fact no J_i can belong even to $\bar{M}\mathfrak{X}$.

Let $I = H \cap K$. We shall construct two descending series

$$H = H_1 \geqq H_2 \geqq \cdots \quad \text{and} \quad K = K_1 \geqq K_2 \geqq \cdots$$

such that $H_{i+1} = A_i^{H_i}$ and $K_{i+1} = B_i^{K_i}$ if $i \geqq 1$, for certain finite subsets A_i of H_i and B_i of K_i, and such that

$$J_i \leqq \langle H_i, K_i \rangle I. \tag{12}$$

When $i = 1$, equation (12) is certainly valid. Suppose that H_i and K_i have been satisfactorily chosen. Then $X_i \subseteq \langle A_i, B_i \rangle I$ for some finite subsets A_i of H_i and B_i of K_i. Let $H_{i+1} = A_i^{H_i}$ and $K_{i+1} = B_i^{K_i}$; then by (11) and (12)

$$J_{i+1} = X_i^{J_i} \leqq \langle A_i^{H_iK}, B_i^{K_iH} \rangle I = \langle H_{i+1}, K_{i+1} \rangle I$$

since $[H, K] \leqq I$. Thus the validity of the construction is assured.

Since H and K belong to $\bar{M}\mathfrak{X}$, there is a positive integer n such that H_n and K_n belong to $\mathfrak{X}$. Hence H_n and K_n are finitely generated. Next $J = HK$, so

$$X_n \subseteq Y_n Z_n \tag{13}$$

for certain finite subsets Y_n of H and Z_n of K. Define

$$P = \langle H_n, K_n, Y_n, Z_n \rangle,$$

a finitely generated subgroup of J. Now $\bar{M}\mathfrak{X} \leqq L\mathfrak{X}$ and $L\mathfrak{X}$ is N_0-closed by the Hirsch-Plotkin-Baer Theorem; therefore $J \in L\mathfrak{X}$ and $P \in \mathfrak{X}$. It follows that P satisfies Max and consequently that Y_n^P and Z_n^P are finitely generated. $Y_n^P \leqq H$ and $Z_n^P \leqq K$ since $H \lhd J$ and $K \lhd J$, and in addition H and K belong to $M^{\alpha}\mathfrak{X}$. Hence Y_n^{PH} and Z_n^{PK} belong to $M^{\beta}\mathfrak{X}$ for some $\beta < \alpha$ since $M^{\beta}\mathfrak{X}$ is evidently S-closed; thus the subgroups

$$U = Y_n^{PI} \quad \text{and} \quad V = Z_n^{PI}$$

belong to $\boldsymbol{M}^{\beta}\mathfrak{X}$. Now U is normalized by P and by I—and also by $\langle Z_n \rangle$ because $Z_n \subseteq P$; hence $U^V = U$ and $U \lhd UV$: for similar reasons $V \lhd UV$. Our induction hypothesis permits us to conclude that $UV \in \overline{\boldsymbol{M}}\mathfrak{X}$. Equations (11), (13) and (12) show that

$$J_{n+1} = X_n^{J_n} \leqq Y_n^{J_n} Z_n^{J_n} \leqq Y_n^{PI} Z_n^{PI} = UV.$$

Finally, it follows that $J_{n+1} \in \overline{\boldsymbol{M}}\mathfrak{X}$, contrary to assumption. □

Corollary. If $\mathfrak{X}$ satisfies the hypothesis of Theorem 2.36, then

$$\overline{\boldsymbol{M}}\mathfrak{X} = \text{Cat } \mathfrak{X} = \boldsymbol{N}\mathfrak{X}.$$

Proof. The definitions assure us that $\overline{\boldsymbol{M}}\mathfrak{X} \leqq \text{Cat } \mathfrak{X} \leqq \boldsymbol{N}\mathfrak{X}$ is valid for any $\mathfrak{X}$. Also $\mathfrak{X} \leqq \overline{\boldsymbol{M}}\mathfrak{X}$ and in our case this leads to

$$\boldsymbol{N}\mathfrak{X} \leqq \boldsymbol{N}(\overline{\boldsymbol{M}}\mathfrak{X}) = \overline{\boldsymbol{M}}\mathfrak{X}. \quad \square$$

Among the possibilities for $\mathfrak{X}$ here are the classes of finite groups and finitely generated nilpotent groups. The second case is of particular interest since it gives a new characterization of Baer groups.

We know from the corollary that

$$\boldsymbol{N}\mathfrak{A} = \boldsymbol{N}(\mathfrak{G} \cap \mathfrak{N}) = \text{Cat } (\mathfrak{G} \cap \mathfrak{N}),$$

and it clear that $\text{Cat } \mathfrak{A} \leqq \text{Cat } (\mathfrak{G} \cap \mathfrak{N}) \leqq \text{Cat } \mathfrak{N}$. Now if $x \in G \in \mathfrak{N}_c$, ($c$ being positive), then $x^G \in \mathfrak{N}_{c-1}$. Thus induction on c yields

$$\mathfrak{N}_c \leqq \mathfrak{A}^{(c)}$$

—the notation is explained on p. 36—and hence

$$\mathfrak{N} \leqq \text{Cat } \mathfrak{A}.$$

Thus $\text{Cat } \mathfrak{N} \leqq \text{Cat } \mathfrak{A}$ and consequently

$$\text{Cat } \mathfrak{A} = \text{Cat } (\mathfrak{G} \cap \mathfrak{N}) = \text{Cat } \mathfrak{N}.$$

It follows that *a Baer group is simply a group with an abelian category* (Rips [1]).

Chapter 3

Maximal and Minimal Conditions

3.1 The Maximal and Minimal Conditions on Subgroups. Polycyclic Groups and Černikov Groups

The general theory of groups which satisfy the maximal condition on subgroups (Max) or the minimal condition on subgroups (Min) is still in its initial stages and is beset with considerable difficulties. Even the minimal condition, generally a more decisive property, is relatively unknown, despite some progress in recent years: this is in contrast with the theory of semisimple rings satisfying the minimal condition on left ideals, which is fairly complete. Thus inevitably we will have mainly to deal here with partial results and special situations.

Polycyclic Groups

Since Max is an extension closed property, every poly-(cyclic or finite) group satisfies Max: indeed the soluble groups which satisfy Max are precisely the polycyclic groups. However *a poly-(cyclic or finite) group is in fact polycyclic-by-finite*. For suppose that $1 = G_m \lhd \cdots \lhd G_1 \lhd G_0 = G$ is a series with cyclic or finite factors in a group G: we can assume that $m > 1$ and that $N = G_1$ is polycyclic-by-finite by induction on m. For some positive integer n the subgroup N^n is polycyclic and N/N^n is finite since it is a finitely generated, periodic, soluble group. If G/N is finite, then G/N^n is finite and G is polycyclic-by-finite. Otherwise G/N is infinite cyclic; in this case let $C = C_G(N/N^n)$. Then G/C is finite, $C/C \cap N$ is cyclic and $C \cap N/C \cap N^n$ is abelian since it is centralized by C. Finally, $C \cap N^n$ is soluble, so C is soluble and, of course, it also satisfies Max. Therefore C is polycyclic and G is polycyclic-by-finite.

No example is known of a group which satisfies Max and is not polycyclic-by-finite, so one is led to ask the question: *is every group which satisfies Max polycyclic-by-finite*? So far very little progress has been made with this problem.

On the other hand, polycyclic groups have been the subject of numerous investigations, beginning with a series of papers by Hirsch [1], [2], [3], [5], [6] in which the basic properties are established. For example, *a polycyclic group has a characteristic series of finite length whose factors are either finite elementary abelian p-groups or free abelian groups of finite rank*; such a series can be obtained by refining the derived series.

Strong Composition and Chief Series in Polycyclic Groups

An infinite polycyclic group does not have either a subnormal composition series or a chief series of finite length since a composition or chief factor of a polycyclic group is finite. However, certain series of a more general type have been introduced by Hirsch [1] in an attempt to obtain analogues of the Jordan-Hölder Theorem. Every polycyclic group has a *weak composition series*, i.e. a series of finite length whose factors are cyclic of prime or infinite order, and also a *weak chief series*, i.e. a normal abelian series of finite length in which each finite factor is a chief factor of G in the usual sense (and so is elementary abelian) and in which each infinite factor has no non-trivial G-admissible subgroups of infinite index (and so is free abelian). By the Schreier Refinement Theorem we see at once that *the infinite factors in a weak composition (chief) series are invariants of G*; but of course the finite factors that can occur in such series may be quite arbitrary since, for example, an infinite cyclic group has a subgroup of every prime index. This leads us to strengthen these concepts. A *strong composition (chief) series* in a polycyclic group G is a weak composition (chief) series of minimal length, which means that the number of *finite* factors is as small as possible. Concerning the isomorphism of such series, there is the following result.

Theorem 3.11 (Zappa [1], p. 11, [2], p. 65). In a supersoluble group any two strong composition (chief) series have isomorphic factors.

We recall that a *supersoluble* group is a group with a normal cyclic series of finite length: thus the class of supersoluble groups is

$$\boldsymbol{P}_n\mathfrak{C}$$

and this is well-known to lie strictly between the class of finitely generated nilpotent groups and the class of polycyclic groups;

$$\mathfrak{G} \cap \mathfrak{N} < \boldsymbol{P}_n\mathfrak{C} < \boldsymbol{P}\mathfrak{C}.$$

In the case of finitely generated nilpotent groups, Theorem 3.11 was proved by Hirsch ([2], Theorem 2.23).

Proof of Theorem 3.11. Let G be a supersoluble group and let $\mathscr{S}$ be a strong composition series of G with length > 1. Let $H \lhd K \lhd L$ be three consecutive terms of $\mathscr{S}$ and suppose that K/H is cyclic of order 2 or ∞, while L/K is cyclic of odd prime order p. Set

$$C = \mathrm{Core}_L\, H = \bigcap_{l \in L} H^l.$$

Then $C \leqq H$, $C \lhd L$ and K/C is either an elementary abelian 2-group or a free abelian group of finite rank. Let $L = \langle t, K \rangle$; then t induces an automorphism of odd order in K/C and, since G is supersoluble, this automorphism may be represented by a triangular matrix with coefficients in the field $GF(2)$ or the ring of integers. Now such a matrix cannot have odd order unless it is the identity. Consequently t centralizes K/C and K/C lies in the centre of L/C. Since $C \leqq H \leqq K$, this implies that $H \lhd L$. The group L/H has cyclic central factor group, so it is abelian. If L/H were torsion-free, it would be infinite cyclic; however this would mean that by omitting K from $\mathscr{S}$ we could obtain a weak composition series of shorter length. Hence L/H has a subgroup of order p, say K^*/H, and K^* is normal in L and L/K^* is cyclic of order 2 or ∞. Now replace the fragment $H \lhd K \lhd L$ of $\mathscr{S}$ by $H \lhd K^* \lhd L$; the new series $\mathscr{S}^*$ has exactly the same factors as $\mathscr{S}$, but a factor of odd order p has been moved one step to the left past a factor of order 2 or ∞. After a finite number of replacements of this type we obtain a series $\mathscr{S}_1$ which is isomorphic with $\mathscr{S}$ but in which all the odd factors occur first. It follows that the elements of odd order in G form a subgroup D and that the odd factors in $\mathscr{S}$ arise from composition factors of D. Thus the factors of odd order in $\mathscr{S}$, as well as the factors of infinite order, are invariants of G. The remaining factors are of order 2 and their number is therefore determined and is an invariant of G. Hence any two strong composition series of G are isomorphic. In a strong chief series of G all factors are cyclic, so the same argument can be applied to show that any two such series are also isomorphic. ▯

On the other hand Bowers [1] has constructed examples of polycyclic groups which have non-isomorphic strong composition and chief-series: in some of these examples the factors of a strong chief-series have rank $\leqq 2$, whereas in a supersoluble group such factors have rank 1. For further results and a discussion of these problems see Hirsch [5] and Bowers [1].

The technique of shifting factors employed in the preceding proof, together with the fact that an infinite cyclic group has automorphism group of order 2, enables us to establish the following: *if G is a supersoluble group, there is a characteristic series $1 \leqq L \leqq M \leqq G$ such that L is finite of odd order, M/L is a finitely generated torsion-free nilpotent group and G/L is a finite 2-group.*

Černikov Groups

A group which is a finite extension of an abelian group satisfying Min is called a *Černikov group*: such groups have also been called *extremal*. Now it is well-known that an abelian group satisfies Min if and only if it is the direct product of a finite number of finite cyclic and *quasicyclic* groups (i.e. groups of type p^∞ for some prime p): see Fuchs [3], p. 65, Theorem 19.2. Hence a Černikov group G contains a subgroup N of finite index such that N is the direct product of finitely many quasicyclic groups. Evidently N is the minimal subgroup of finite index in G, so it is characteristic in G.

Every Černikov group satisfies Min and is locally finite and countable since these properties are extension closed; but *it is an open question whether every group which satisfies Min is a Černikov group or locally finite or even countable.* Since every group which satisfies Min is periodic, the second of these questions is a special type of Burnside problem.

Soluble Groups Satisfying Min

The structure of a soluble group satisfying Min was determined originally by Černikov ([4], Theorems 3 and 4), who showed that such a group is what we have called a Černikov group. More generally there is

Theorem 3.12 (cf. Kuroš [9], Vol. 2, § 59). An SI-group which satisfies Min-*sn* (the minimal condition on subnormal subgroups) is a soluble Černikov group.*

Proof. Let G be a SI-group which satisfies Min-*sn*. By definition G has a normal abelian series and by Min-*sn* this is an ascending series; in other words G is hyperabelian.

Let R denote the finite residual of G; then G/R is finite in view of the property Min-*sn*. If R has a subgroup H with finite index, then H has finite index in G; hence $N_G(H)$ — and therefore $\mathrm{Core}_G\, H$ — has finite index. It follows that $R \leqq \mathrm{Core}_G\, H \leqq H$, so $H = R$. Thus R has no proper subgroups of finite index. If $R = 1$, then G is finite, so it is certainly a Černikov group; let us assume that $R \neq 1$.

Since G is hyperabelian, R contains a non-trivial normal abelian subgroup of G, say A, by Lemma 2.16. Let $1 \neq a \in A$; now G is periodic, so a has finite order and every conjugate of a in G lies in A and has the same finite order as a. But $A \; sn \; G$, so A satisfies Min and, by the structure of abelian groups with the property Min, the number of elements in A with order equal to that of a is finite. Hence $|R: C_R(a)|$ is finite and consequently $R = C_R(a)$. This means that $a \in \zeta(R)$ and $\zeta(R) \neq 1$. If $R = \zeta(R)$, then R is an abelian group satisfying Min and G is a Černikov group. Suppose therefore that $\zeta(R) < R$. Now $\zeta(R) \lhd G$, so by the previous argument $\zeta(R/\zeta(R))$ is non-trivial and $\zeta(R) = \zeta_1(R) < \zeta_2(R)$. Let $z \in \zeta_2(R) \backslash \zeta(R)$ and let $x \in R$. Then $z^x = zz_1$ where $z_1 \in \zeta(R)$. Since z and z_1 commute, it follows that the order of z_1 divides that of z. Hence z

* See also Theorem 5.46.

has only a finite number of conjugates in R and $|R: C_R(z)|$ is finite. But R has no proper subgroups of finite index, so $z \in \zeta(R)$. By this contradiction G is a Černikov group. Finally, G/R is a finite hyperabelian group, so it—and therefore G—is soluble. □

Corollary. The class of Černikov groups is $\boldsymbol{P}$-closed and is precisely the class of poly-(quasicyclic or finite) groups.

Proof. It is enough to prove that the class of Černikov groups is $\boldsymbol{P}$-closed. Let $N \lhd G$ where N and G/N are Černikov groups and let R and S/N be the finite residuals of N and G/N respectively. Then $S \lhd G$ and $R \lhd G$ since R is characteristic in N. Also G/S and N/R are finite, while R and S/N are abelian. Let C be the centralizer of N/R in S. Clearly $C \lhd G$ and G/C is finite, while $1 \leqq R \leqq C \cap N \leqq C$ is an abelian series in C, so that C is soluble. But G, and hence C, satisfies Min, by $\boldsymbol{P}$-closure, and Theorem 3.12 shows that C is a Černikov group. Finally, G/C is finite, so G is a Černikov group. □

Theorem 3.12 shows that the soluble groups which satisfy Min are precisely the soluble Černikov groups and that they are also just the poly-(finite cyclic or quasicyclic) groups.

Let us now determine the structure of nilpotent groups which satisfy Min. First we need some terminology and a lemma. A group is said to be *radicable* if each element is an nth power for every positive integer n.

Lemma 3.13. Let A be a normal, radicable, abelian subgroup of a group G and let H be a subgroup of G such that

$$[A, \underbrace{H, \ldots, H}_{r}] = 1$$

for some positive integer r. If H/H' is periodic, then $[A, H] = 1$.

Proof. Define $A_0 = A$ and $A_{i+1} = [A_i, H]$ if $i \geqq 0$; then $A_r = 1$. If $a \in A$ and $h \in H$, the mapping

$$(a, hH') \to [a, h] A_2$$

is well-defined because $[A, H'] \leqq [A, H, H] = A_2$ and bilinear because A is abelian. Hence there is induced a homomorphism of $A \otimes (H/H')$ onto A_1/A_2. But A is radicable and H/H' is periodic, so $A \otimes (H/H')$ is trivial; therefore $A_1 = A_2$. This implies that $A_1 = A_2 = \cdots = A_r$, so that $A_1 = 1$ as required. □

Theorem 3.14. (Baer [20], § 6, gatz 2). G is a nilpotent group satisfying Min if and only if $\zeta(G)$ satisfies Min and $G/\zeta(G)$ is a finite nilpotent group.

Proof. Let G be nilpotent group satisfying Min. By Theorem 3.12 G is a Černikov group and there is a normal, radicable, abelian subgroup R

with finite index in G. Let $g \in G$: since G is nilpotent,

$$[R, \underbrace{\langle g \rangle, \ldots, \langle g \rangle}_{c}] = 1$$

where c is the nilpotent class of G. The group G is periodic, so Lemma 3.13 shows that $[R, g] = 1$ and $R \leq \zeta(G)$. Hence $|G: \zeta(G)|$ is finite. The converse is obvious. □

From this it is clear that the nilpotent groups which satisfy Min are precisely those Černikov groups whose finite residual lies in the centre and has nilpotent factor group.

2-groups which Satisfy Max or Min

The structure of p-groups which satisfy Max or Min (where p is a prime) is still unknown except when $p = 2$. In this case the two conjectures about Max and Min are true, but of course this is scant evidence for the case of a general prime p.

Theorem 3.15. In an infinite 2-group every finite subgroup is properly contained in its normalizer.

Proof. Let F be a finite subgroup of an infinite 2-group G and assume that $F = N_G(F)$. Since G is infinite, not all its finite subgroups are contained in F; hence we can choose a finite subgroup M not contained in F such that $I = M \cap F$ is maximal. Since $M \nleq F$, we have $I < M$. Now in a nilpotent group every subgroup is subnormal and hence every proper subgroup is properly contained in its normalizer (see p. 49). Therefore $I < N_M(I)$. Also $F = N_G(F)$, so $I < F$ and $I < N_F(I)$.

Consequently we can find elements $x \in M$ and $y \in F$ such that $I^x = I = I^y$ and xI and yI are *involutions* (i.e. elements of order 2). Let $T = \langle x, y, I \rangle$. Now $I \lhd T$ and T/I is generated by two involutions, so it is a periodic homomorphic image of the free product of a pair of infinite cyclic groups. Such a free product $P = \langle u \rangle * \langle v \rangle$ is an infinite dihedral group: for if $w = uv$, then $P = \langle u, w \rangle$ and $w^u = w^{-1}$. Hence T/I is finite and consequently T is finite. But $T \nleq F$ since $x \notin I$, and $I < T \cap F$ since $y \notin I$. This contradicts the maximality of I. □

Corollary. (Kegel [4]). A 2-group which satisfies the maximal condition on finite subgroups is finite; in particular a 2-group which satisfies Max is finite.

To study 2-groups which satisfy Min it is necessary to know the structure of locally finite 2-groups satisfying Min.

Theorem 3.16. A locally nilpotent group which satisfies Min is a hypercentral Černikov group.

We prefer not to give a proof at this stage since the result is a special case of a more general theorem (Theorem 5.27, Corollary 2; see also part (ii) of the proof of Theorem 3.32).

Notice that Theorem 3.16 permits us to determine the SN-groups which satisfy Min. For if G is such a group, it is clearly an SN^*-group, and therefore a radical group—see Section 2.3. Theorem 3.16 shows that G is hyperabelian and Theorem 3.12 that G is a soluble Černikov group. This we state as a

Corollary (O. J. Schmidt [5], Theorems 7 and 8). An SN-group which satisfies Min is a soluble Černikov group.

Theorem 3.17 (O. J. Schmidt [6], pp. 298—300, and [7]). A 2-group which satisfies Min is a Černikov group and hence is locally finite, countable and soluble.

Proof. Suppose that there exists a group—not necessarily a 2-group—which satisfies Min yet is not locally finite. Then it contains a subgroup G which is minimal with respect to not being locally finite; thus every proper subgroup of G, but not G itself, is locally finite. The union of a chain of locally finite groups is always locally finite, so G cannot be such a union and therefore each proper subgroup (proper normal subgroup) of G lies in a maximal subgroup (maximal normal subgroup) of G. Let N be a maximal normal subgroup of G and set $H = G/N$. Then H is a simple non-locally finite group satisfying Min, with locally finite proper subgroups; for if H were locally finite, so would G be, by the $\boldsymbol{P}$-closure of the class $\boldsymbol{L}\mathfrak{F}$ (Theorem 1.45). Notice also that H is finitely generated since otherwise finitely generated subgroups of H would be proper and therefore finite.

From now on we assume that G—and hence H—is a p-group for some prime p and we will show that *each pair of distinct maximal subgroups of H intersect trivially*. Suppose that this is false and let M and M_1 be distinct maximal subgroups of H such that

$$I_1 = M \cap M_1 \neq 1.$$

In the ensuing discussion it is understood that we have a fixed M in mind. M_1 is a locally finite p-group and hence is locally nilpotent; it also satisfies Min, so it is hypercentral by Theorem 3.16. Now every subgroup of a hypercentral group is ascendant (see pp. 49—50); also $I_1 < M_1$. Therefore I_1 is a proper subgroup of its normalizer in M_1. Since Z_1 the centre of I_1, is characteristic in I_1, we see that Z_1 must be normalized by an element of M_1 lying outside M. Also $Z_1 \neq 1$ and H is simple, from which we deduce that $N_H(Z_1) < H$. Hence there is a maximal subgroup M_2 of H containing $N_H(Z_1)$. Certainly $M \neq M_2$ since $N_H(Z_1) \nleq M$. Also $I_1 \leqq N_M(Z_1) \leqq M_2$, so that $I_1 \leqq I_2 = M \cap M_2$.

Since H satisfies Min, we may suppose that M_1 and M_2 have been chosen subject to the *additional* requirement that $Z_2 = \zeta(I_2)$ be minimal. Just as before, we obtain $N_H(Z_2) < H$ and $N_H(Z_2) \nleq M$. Thus it is possible to find a maximal subgroup M_3 of H which contains $N_H(Z_2)$ and is distinct from M. Now

$$I_1 \leqq I_2 < N_M(I_2) \leqq N_M(Z_2) \leqq M \cap M_3 = I_3,$$

say. Therefore $Z_3 = \zeta(I_3)$ centralizes Z_1 and consequently

$$Z_3 \leqq N_M(Z_1) \leqq M \cap M_2 = I_2.$$

Hence $Z_3 \leqq \zeta(I_2) = Z_2$. Now the pair (M_2, M_3) has all the attributes of the pair (M_1, M_2). Hence we must have $Z_2 = Z_3$: therefore

$$I_3 \leqq N_M(Z_3) = N_M(Z_2) \leqq M \cap M_3 = I_3.$$

It follows that $I_3 = N_M(Z_3)$. However $I_3 < M$, which implies that Z_3 is normalized by an element of M lying outside M_3, a contradiction which establishes our original assertion $M \cap M_1 = 1$.

In the final part of the proof we assume that $p = 2$. Let M and M_1 be two distinct maximal subgroups of H: these exist because H is simple. Let a and a_1 be elements of order 2 in M and M_1 respectively. Since $\langle a, a_1 \rangle$ is a dihedral group, it is soluble and it is therefore a proper subgroup of H. Let M_2 be a maximal subgroup of H containing $\langle a, a_1 \rangle$. Then $M \cap M_2 \neq 1$ and $M_1 \cap M_2 \neq 1$, so by the last paragraph $M = M_2 = M_1$.

So far we have proved that a 2-group which satisfies Min is locally finite. Consequently, such a group is locally nilpotent, and Theorem 3.16 now implies that it is a Černikov group. □

Both Theorem 3.15 and Theorem 3.17 depend ultimately on the fact that for $p = 2$ two elements of order p in a group always generate a soluble subgroup. However, if p is odd, the argument breaks down completely at this point (cf. the theorem of P. Hall mentioned on p. 30).

Further cases when Min implies that a group is a Černikov group are discussed in Section 3.4. See also Baer [34].

A Criterion of Baer for a Group to Satisfy Max Locally

A group satisfies Max locally if and only if its finitely generated subgroups satisfy Max. Recall that the groups which satisfy Max locally form an $\acute{\boldsymbol{N}}$-closed class, by Theorem 2.31: however this class is not a radical class since it is not $\boldsymbol{P}$-closed, as may be seen from the group with generators a and b and the single relation $b^{-1}ab = a^2$. In view of this example the following result is of interest.

Theorem 3.18 (Baer [26], Satz 5). A group which has an ascending normal series whose factors are finitely presented and satisfy Max must itself satisfy Max locally.

Proof. Let G be a group satisfying the hypothesis of the theorem. Denote by $\mathscr{S}$ the set of normal subgroups N of G with the following property: if $N \leqq X \leqq G$ and X/N is finitely presented and satisfies Max, then X satisfies Max locally. $\mathscr{S}$ is not empty since it contains the trivial subgroup. Let $\{N_\alpha : \alpha \in A\}$ be a chain in $\mathscr{S}$, ordered by inclusion, and let U be its union: we propose to show that $U \in \mathscr{S}$. Assume that $U \leqq X \leqq G$ and that X/U is finitely presented and satisfies Max. Let H be a finitely generated subgroup of X; we have to show that H satisfies Max and there is nothing lost in assuming that $X = HU$ here because X/U is finitely generated. Since $X/U \simeq H/H \cap U$, the latter is finitely presented and consequently $H \cap U = x_1^H \cdots x_n^H$ for some finite subset $\{x_1, \ldots, x_n\}$ of $H \cap U$: this is by Lemma 1.43. Now for some $\alpha \in A$, all the x_i are contained in N_α and, since $N_\alpha \lhd G$, this implies that

$$H \cap U = H \cap N_\alpha .$$

Therefore $HN_\alpha/N_\alpha \simeq H/H \cap U$, which is finitely presented and satisfies Max. But $N_\alpha \in \mathscr{S}$, so HN_α satisfies Max locally by definition of $\mathscr{S}$, and H, being finitely generated, satisfies Max. Hence $U \in \mathscr{S}$. Zorn's Lemma may now be applied to obtain a maximal element of $\mathscr{S}$, say M. If $M = G$, the defining property of $\mathscr{S}$ implies that G satisfies Max locally. Suppose that $M \neq G$. The group G/M cannot have a non-trivial normal subgroup which is finitely presented and which satisfies Max: for, by the $\boldsymbol{P}$-closure of the properties finitely presented and Max (Lemma 1.43 and Corollary, Lemma 1.48), such a subgroup would give rise to a member of $\mathscr{S}$ properly containing M. Now observe that the class of finitely presented groups which satisfy Max is $\boldsymbol{H}$-closed; then Lemma 1.22 implies that G/M is trivial. ▯

By Lemma 1.43 polycyclic groups are finitely presented, so Theorem 3.18 has the following

Corollary. A hyper-(finite or polycyclic) group satisfies Max locally.

A further special case is the result of Baer that *hypercyclic groups satisfy Max locally and hence are locally supersoluble* (Baer [21], Theorem 1).

It is unknown whether a group with an ascending normal series all of whose factors satisfy Max need itself satisfy Max locally. Certainly this would be the case if every group with Max were polycyclic-by-finite.

For related results see Plotkin [12] (§ 18) and Simon [1].

Some Related Finiteness Conditions

Maximal and minimal conditions on abelian, subnormal and normal subgroups are considered in detail in later sections. We mention briefly at this point some weak forms of the minimal condition which will not be discussed here but which have received attention. The *minimal condition on p-subgroups* has been studied by Černikov ([26], § 8) – see also Wehrfritz [8]. Polovickiĭ [1], [4] has investigated a similar property, the *π-minimal condition* where π is a set of primes: this excludes the possibility of an infinite descending chain of subgroups $G_1 > G_2 > \cdots$ such that each set $G_i \backslash G_{i+1}$ contains at least one element of order a π-number. In [4], [6] and [8] Zaičev studies groups in which there are no infinite descending chains of subgroups $G_1 > G_2 > \cdots$ such that each index $|G_i : G_{i+1}|$ is infinite, and in [1] groups which have no infinite descending chains of subgroups $G_1 > G_2 > \cdots$ such that the sets of commutators in the G_i steadily decrease.

3.2 Groups of Automorphisms of Soluble Groups. Soluble Linear Groups

In the theory of soluble groups, groups of automorphisms of soluble groups, and particularly of abelian groups, may be expected to play a major role. For example, when we come to discuss the properties Max-*ab* and Min-*ab* in the following section, it will be vital to have information about groups of automorphisms of finitely generated abelian groups and of abelian groups which satisfy Min. Of course, a linear group can also be regarded as a group of automorphisms of an abelian group: we will begin by considering linear groups.

Soluble Linear Groups

The theory of soluble linear groups is in a well-developed state—see for example the book by Suprunenko [1]—and we present only those results that are essential for our investigations.

First of all some terminology: a group G of linear transformations of a finite dimensional vector space V is said to be *triangulable*, *unitriangulable* or *diagonable* if the elements of G can be simultaneously represented by (upper) triangular, (upper) unitriangular or diagonal matrices with respect to a suitable basis of V. It is easy to verify that a diagonable group is abelian and that a unitriangulable group of positive degree n is nilpotent of class $\leqq n - 1$; thus a triangulable group is nilpotent-by-abelian. It is of interest to know how far the converses of these statements are true. A partial answer is furnished by the following fundamental theorem.

Theorem 3.21 (Mal'cev [7], Theorem 1). Let G be a soluble group of linear transformations of an n-dimensional vector space V over an algebraically closed field F. Then G has a normal triangulable subgroup with finite index not exceeding

$$\prod_{i=1}^{n} (i!)\,(i^2 f(i^2))^i$$

where $f(i)$ is the maximum number of automorphisms of a group of order $\leqq i$. Furthermore, if G is irreducible, it has a normal diagonable subgroup with finite index not exceeding

$$n!\,(n^2 f(n^2))^n.$$

The proof below is a simplification due to B. A. F. Wehrfritz of that in Suprunenko's book ([1], Theorem 15). The main step in the proof is

Lemma 3.22. Let G be as in Theorem 3.21 and suppose that G is primitive and irreducible. Then the centre of G consists of scalar linear transformations and has finite index in G not exceeding $n^2 f(n^2)$.

Proof. Let A be a normal abelian subgroup of G: as usual FA denotes the group algebra of A over F. By Clifford's Theorem we can write

$$V = V_1 \oplus \cdots \oplus V_k$$

where each V_i is the sum of all the irreducible FA-submodules of V isomorphic with a given one. Since A is abelian and F is algebraically closed, each irreducible FA-module has dimension 1 and to each V_i there corresponds a characteristic function $\xi_i\colon A \to F$ such that if $v \in V_i$ and $a \in A$,

$$va = \xi_i(a)\,v.$$

In addition the ξ_i are all distinct. Let i be a fixed integer such that $1 \leqq i \leqq k$, let $v \in V_i$ and let $g \in G$. Then

$$vg = \sum_{j=1}^{k} v_j$$

for certain $v_j \in V_j$. If $a \in A$, we have

$$(vg)\,a = \sum_{j=1}^{k} v_j a = \sum_{j=1}^{k} \xi_j(a)\,v_j$$

and also, since $A \lhd G$,

$$(vg)\,a = (va^{g^{-1}})\,g = \xi_i(a^{g^{-1}})\,vg = \sum_{j=1}^{k} \xi_i(a^{g^{-1}})\,v_j.$$

Since the V_j are independent, either $v_j = 0$ or $\xi_j(a) = \xi_i(a^{g^{-1}})$ for all $a \in A$. But the ξ_j are distinct, so there is at most one j such that $v_j \neq 0$ and therefore $vg = v_j \in V_j$. It follows that $V_ig \leq V_j$; also $V_jg^{-1} \leq V_i$ in the same way, so $V_ig = V_j$ and the elements of G permute the subspaces $V_1, \dots, V_k$. By definition of primitivity, $k = 1$ and $V = V_1$. This means that A consists of scalar linear transformations; therefore A is contained in Z, the centre of G, and every normal abelian subgroup of G lies in Z.

Now let B/Z be a maximal normal abelian subgroup of G/Z: our immediate aim is to prove that

$$|B:Z| \leq n^2 \quad \text{and} \quad C_G(B/Z) = B. \tag{1}$$

From these facts we will be able to conclude that

$$|G:B| \leq |\mathrm{Aut}(B/Z)| \leq f(n^2)$$

and hence that $|G:Z| \leq |G:B|\,|B:Z| \leq n^2 f(n^2)$, as required.

Suppose first of all that $C_G(B) \nleq Z$. Since G/Z is a soluble group, $C_G(B)/Z$ contains a non-trivial normal abelian subgroups of G/Z, say C/Z. Now BC/Z is abelian because $[B, C] = 1$, so by maximality of B/Z we have $C \leq B \cap C_G(B) = \zeta(B)$. Since Z contains all normal abelian subgroups of G, it follows that $C \leq Z$. This contradicts the choice of C, so

$$C_G(B) = Z.$$

Next let $\{b_1, \dots, b_r\}$ be a subset of a transversal to Z in B. Suppose that this set is linearly dependent over F: then, after relabelling the b_i if necessary, we can find a relation

$$\sum_{i=1}^{s} \lambda_i b_i = 0$$

where the λ_i are non-zero scalars and the length s is minimal. Since $b_1 b_2^{-1} \notin Z = C_G(B)$, there is an $x \in B$ such that $[b_1 b_2^{-1}, x] \neq 1$. Now $[b_i, x] \in Z$ since B/Z is abelian; consequently

$$[b_1 b_2^{-1}, x] = [b_1, x]\,[b_2, x]^{-1}$$

and $[b_1, x] \neq [b_2, x]$. Now Z is scalar and $[b_i, x] = \beta_i 1$ where $\beta_i \in F$. Then $\beta_1 \neq \beta_2$ and

$$0 = \beta_1 \left(\sum_{i=1}^{s} \lambda_i b_i\right) - x^{-1} \left(\sum_{i=1}^{s} \lambda_i b_i\right) x = \sum_{i=2}^{s} (\beta_1 - \beta_i)\,\lambda_i b_i.$$

Since $(\beta_1 - \beta_2)\,\lambda_1 \neq 0$, this contradicts the minimality of s. Hence $b_1, \dots, b_r$ are linearly independent and $|B:Z|$ is finite and cannot exceed n^2, the dimension of the vector space of all linear transformations of V.

It remains to show that $C_G(B/Z) = B$. Let $c \in C_G(B/Z)$. Then the map θ_c: $bZ \to [b, c]$ is a homomorphism of B/Z into Z and $c \to \theta_c$ is a homomorphism of $C_G(B/Z)$ into $\mathrm{Hom}(B/Z, Z)$ with kernel equal to

$C_G(B)$, that is to Z. This shows that $(C_G(B/Z))/Z$ is isomorphic with a subgroup of $\mathrm{Hom}(B/Z, Z)$. Since Z is scalar, it is isomorphic with a subgroup of the multiplicative group of the field F; therefore, finite subgroups of Z are cyclic. Hence

$$|\mathrm{Hom}(B/Z, Z)| \leqq |B:Z|$$

and therefore

$$|C_G(B/Z):Z| \leqq |B:Z|.$$

But it is clear that $B \leqq C_G(B/Z)$, so $B = C_G(B/Z)$ by comparison of orders. ▯

Proof of Theorem 3.21. By induction on n it is sufficient to prove the second part of the theorem, when G is irreducible. If G is primitive, $|G: \zeta(G)| \leqq n^2 f(n^2)$, while $\zeta(G)$ is diagonable (and even scalar) by Lemma 3.22. Suppose therefore that G is imprimitive, which means that there is a decomposition of V into non-zero subspaces

$$V = V_1 \oplus \cdots \oplus V_k$$

where $k > 1$ and the elements of G permute the V_i. We can assume in addition that k is maximal with this property. An element g of G induces a permutation π_g of the V_i and, naturally, the mapping $g \to \pi_g$ is a homomorphism of G into the symmetric group of degree k: let K be the kernel of this homomorphism. Then K is the intersection of all the subgroups $S_i = \{g: g \in G, V_i g = V_i\}$. Now S_i acts on V_i as a group of linear transformations $\bar{S}_i$, and it is straightforward to show that $\bar{S}_i$ is primitive by the maximality of k and that $\bar{S}_i$ is irreducible by the irreducibility of G. Lemma 3.22 applies and we conclude that S_i has a normal subgroup D_i of index at most $n_i^2 f(n_i^2)$, where $n_i = \dim V_i$ and D_i acts as a group of scalar linear transformations of V_i, say $\bar{D}_i$. Let D be the intersection of all the D_i. Then D is diagonable as a group of linear transformations of V while $D \leqq K$ and

$$|G: D| = |G: K|\,|K: D| \leqq (n!)\,(n^2 f(n^2))^n.$$

It is clear from Lemma 3.22 that we can choose $\bar{D}_i$ to be the centre of $\bar{S}_i$, so that elements of G permute the D_i in the same way as the V_i. Hence $D \lhd G$ and the proof is complete. ▯

A rough upper bound for $f(m)$ is

$$(m-1)(m-2)(m-2^2)\ldots, \quad (m > 1),$$

While more precise results than this are available, it seems unlikely that a bound for the index of a triangulable subgroup obtained in this manner will be sharp. However, the mere existence of a bound is sufficient for the proof of the following result of Zassenhaus.

Theorem 3.23 (Zassenhaus [1], Satz 7). The derived length of a soluble linear group of degree n over an arbitrary field does not exceed a certain number depending only on n

Proof. Let G be a soluble group of linear transformations of degree n over a field F. Without loss of generality we can assume that F is algebraically closed. Then by Theorem 3.21 the group G has a normal triangulable subgroup T of index not exceeding $g(n)$, the first bound specified in that theorem. G/T has derived length at most $[\log_2 g(n)]$ and, if $n > 1$, a group of $n \times n$ upper triangular matrices has derived length at most $[\log_2 (n-1)] + 2$. Hence the derived length of G is at most $[\log_2 g(n)] + [\log_2 (n-1)] + 2$. ▯

Corollary. A linear group belonging to the class $\langle \acute{\mathbf{P}}, \mathbf{L} \rangle \mathfrak{A}$ is soluble.

Proof. Let G be such a group. It is easy to verify that the union of all the classes $(\mathbf{L}\acute{\mathbf{P}})^{\alpha} \mathfrak{A}$ for α an ordinal is both $\mathbf{L}$ and $\acute{\mathbf{P}}$-closed, from which it follows that this union is precisely $\langle \acute{\mathbf{P}}, \mathbf{L} \rangle \mathfrak{A}$. Hence there is an ordinal α for which

$$G \in (\mathbf{L}\acute{\mathbf{P}})^{\alpha+1} \mathfrak{A} = (\mathbf{L}\acute{\mathbf{P}}) (\mathbf{L}\acute{\mathbf{P}})^{\alpha} \mathfrak{A},$$

and by induction on α we obtain $G \in \mathbf{L}\acute{\mathbf{P}}(\acute{\mathbf{P}}\mathfrak{A}) = \mathbf{L}\acute{\mathbf{P}}\mathfrak{A}$. Zassenhaus' theorem shows immediately that a locally soluble, linear group is soluble, so we can assume that $G \in \acute{\mathbf{P}}\mathfrak{A}$, i.e. G is an SN^*-group. Let $\{G_\beta\}$ be an ascending abelian series of G and let β be the first ordinal such that G_β is not soluble. Obviously β must be a limit ordinal; but this implies that G_β is locally soluble since it is a union of soluble subgroups; hence G_β is soluble. ▯

Thus in particular a *radical linear group is soluble* (Plotkin [7], Theorem 13).

Not surprisingly, the bounds for the derived length of a soluble linear group of degree n arising from our proof of Zassenhaus' theorem are of quite the wrong order. For more precise bounds see Huppert [1] and Dixon [4]. In another paper [3] Dixon has obtained upper bounds for the index of the Fitting subgroup of a completely reducible, soluble linear group over an algebraically closed field.

Theorem 3.24. A soluble linear group is nilpotent-by-abelian-by-finite and an irreducible soluble linear group is abelian-by-finite.

This is a consequence of Theorem 3.21. More generally, part one of Theorem 3.24 is true for soluble groups of automorphisms of a finitely generated module over a commutative Noetherian ring (Gruenberg [6], p. 296).

A soluble group which is not in the class $\mathfrak{NAF}$ cannot, therefore, be linear over any field. For example, the standard wreath product $(A \wr B) \wr C$ of three infinite cyclic groups is a finitely generated soluble group that is not linear.

Groups of Automorphisms and Linear Groups

Let C be an additive abelian group and let A be its group of automorphisms. In certain circumstances it is possible to regard A as a linear group over an integral domain, the most obvious case being when C is an elementary abelian p-group. We shall consider two further cases.

Suppose that C is a direct sum of n groups of type p^∞, say $C_1, \ldots, C_n$. Since $\mathrm{Hom}(C_i, C_j)$ is isomorphic with the ring of p-adic integers R_p, the automorphism group A is isomorphic with $GL(n, R_p)$, the group of all $n \times n$ matrices over R_p with determinant equal to a unit of R_p; see Fuchs [3] (p. 212, Theorem 55.1).

Now let C be a torsion-free abelian group of finite rank n. If Q denotes the field of rational numbers, the tensor product $V = C \otimes Q$ (over the ring of integers) is a vector space of dimension n over Q: moreover, the mapping $c \to c \otimes 1$ is an isomorphism of C with a subgroup of V. Let α belong to A; then α can be extended to a non-singular linear transformation α^* of V defined by

$$(c \otimes \lambda)\,\alpha^* = (c\alpha) \otimes \lambda, \quad (c \in C, \lambda \in Q).$$

The mapping $\alpha \to \alpha^*$ determines an isomorphism of A with a subgroup of $GL(n, Q)$, the group of non-singular $n \times n$ matrices over Q.

We will use these ideas to apply Mal'cev's theorem on soluble linear groups. The value of this theorem in the theory of abstract soluble groups is illustrated by the next result.

Theorem 3.25 (Mal'cev [7], Theorem 4). Let G be a group which has an abelian series of finite length such that if F is any factor of the series, the torsion-subgroup T of F satisfies Min and F/T is a torsion-free abelian group of finite rank. Then G is nilpotent-by-abelian-by-finite.

Proof. Let $1 = G_n \lhd G_{n-1} \lhd \cdots G_1 \lhd G_0 = G$ be the given series in G. The first point to be made is that there is a normal (even characteristic) series in G with the same type of factors and with length $\leq n$. Let

$$H = \bigcap_{\alpha \in \mathrm{Aut}\, G} G_1^\alpha$$

and observe that H is characteristic in G; also $H \leq G_1$ and G/H is abelian since $G' \leq G_1^\alpha$ for all $\alpha \in \text{Aut } G$. Let T/H be the torsion-subgroup of G/H. Then it is clear that G/T is torsion-free and abelian of finite rank and that, for each prime p, the p-component of G/H satisfies Min (for a primary homomorphic image of a torsion-free abelian group of finite rank satisfies Min). To prove that T/H satisfies Min we need only show that the set of primes dividing the orders of elements of G/H is finite. Let xH have prime order p: then $x \notin G_1^\alpha$ for some $\alpha \in \text{Aut } G$, while $x^p \in H \leq G_1^\alpha$. Therefore xG_1^α has order p: but $G/G_1^\alpha \simeq G/G_1$ and hence there are only finitely many such primes p. Thus we can take H to be the second term of our new series; since $H \leq G_1$, induction on n allows us to complete the series down to 1 in at most n steps. This part of the argument is due to Gruenberg ([4], Lemma 5.2).

Assume now that $1 = G_n \leq G_{n-1} \leq \cdots G_1 \leq G_0 = G$ is a *normal* series with factors of the type described in the theorem. In view of the structure of abelian groups with the property Min, we can assume, at the expense of prolonging the series by a finite amount, that each factor is either a torsion-free abelian group of finite rank, a finite elementary abelian p-group or a direct product of finitely many groups of type p^∞. Let F be a factor of the series; then the remarks which precede this proof show that $G/C_G(F)$ can be identified with a group of linear transformations of a finite dimensional vector space over the field of rational numbers, the field of p elements $GF(p)$ or the field of p-adic numbers, for some prime p. Passing to the algebraic closure of the field concerned and applying Mal'cev's theorem, we conclude that $G/C_G(F)$ has a normal subgroup of finite index whose derived group is unitriangulable. It follows that G has a normal subgroup N with finite index such that

$$[G, \underbrace{N', \ldots, N'}_{m}] = 1$$

for some integer m. Clearly N' is nilpotent and $G \in \mathfrak{N}\mathfrak{A}\mathfrak{F}$. (Notice that the index $|G:N|$ is bounded by a number depending only on the length of the series and the ranks of the factors.) $\square$

Corollary. Polycyclic groups are nilpotent-by-abelian-by-finite.

Irreducible and Rationally Irreducible Groups of Automorphisms

Let A be a group of automorphisms of an abelian group C. Then A is said to be *irreducible* if there are no proper non-trivial A-admissible subgroups of C. More generally, A is *rationally irreducible* if C/B is periodic whenever B is a non-trivial A-admissible subgroup of C.

Let C be a torsion-free abelian group and let $A \leq \text{Aut } C$. We extend the action of A to the rational vector space $V = A \otimes Q$ as explained above: here Q is the field of rational numbers. Then *A is rationally irreducible on C if and only if A is irreducible as a group of linear transformations of V*. We shall find frequent use for this easily established fact.

Abelian Subgroups of the Unimodular Group

Let Z denote the ring of rational integers and let

$$GL(n, Z)$$

denote the group of all $n \times n$ matrices over Z with determinant ± 1, i.e. the *unimodular group* of degree n.

Lemma 3.26. Abelian subgroups of $GL(n, Z)$ are finitely generated, that is to say, $GL(n, Z)$ satisfies Max-*ab*.

Proof. (P. Hall [6], pp. 71–72). Let A be an abelian subgroup of $GL(n, Z)$, so that A can be regarded as a group of automorphisms of a free abelian group of rank n. Suppose first of all that A is rationally reducible; then $n > 1$ and elements of A can be transformed simultaneously into the form

$$x = \begin{pmatrix} \alpha(x) & \gamma(x) \\ 0 & \beta(x) \end{pmatrix}$$

where $\alpha(x) \in GL(r, Z)$, $\beta(x) \in GL(n - r, Z)$, $\gamma(x)$ is an $r \times (n - r)$ matrix over Z and $0 < r < n$. Those x in A for which $\alpha(x)$ and $\beta(x)$ are trivial form a normal subgroup U of G which is isomorphic with a subgroup of the *additive* group of all $r \times (n - r)$ matrices over Z. Since the latter is a free abelian group of rank $r(n - r)$, the subgroup U is finitely generated. But A/U is isomorphic with an abelian subgroup of the direct product of $GL(r, Z)$ and $GL(n - r, Z)$. We can now use induction on n to conclude that A/U, and hence A, is finitely generated.

We can therefore assume that A is rationally irreducible. Let $\mathscr{C}$ be the centralizer of A in the ring of all $n \times n$ matrices whose coefficients are rational numbers. Then $\mathscr{C}$ is a finite dimensional algebra over the field of rational numbers and by Schur's Lemma $\mathscr{C}$ is a division algebra (see Kuroš [12], p. 217). Hence $\mathscr{D}$, the centre of $\mathscr{C}$, is a finite field extension of the field of rational numbers, i.e. it is an algebraic number field. Also $A \leqq \mathscr{D}$. Let $a \in A$: since $\det a = \det (a^{-1}) = \pm 1$, the characteristic polynomials of a and a^{-1} have constant terms equal to ± 1 and the Cayley-Hamilton Theorem shows that a is an algebraic unit of $\mathscr{D}$. Hence A is a subgroup of the multiplicative group of units of $\mathscr{D}$. But a classical theorem of Dirichlet asserts that the group of units of $\mathscr{D}$ is finitely generated (Hasse [1], p. 399). Hence A is finitely generated. ▯

In his important paper of 1951 Mal'cev proved that a soluble group of automorphisms of a finitely generated abelian group is polycyclic (Mal'cev [7], Theorem 2). Subsequently this was generalized by Smirnov ([4], Theorem 3) and Baer ([24], Satz B'). All of these results are special cases of the following theorem.

Theorem 3.27. A radical group of automorphisms of a polycyclic-by-finite group is polycyclic.

Proof. Assume first that A is an abelian group of automorphisms of a polycyclic-by-finite group G; we show that A is finitely generated. By taking a suitable power of G we obtain a characteristic polycyclic subgroup of finite index in G. By elementary properties of polycyclic groups, G has a characteristic series of finite length whose factors are either finite or free abelian of finite rank. Let N be the least non-trivial term of this series. Then, by induction on the length of the series and by Lemma 3.26, A induces finitely generated groups of automorphisms in G/N and N. Thus, if B is the subgroup of all elements of A that act trivially on G/N and N, the group A/B is finitely generated. Let $g_1, \ldots, g_n$ be a finite set of generators for G; the map $\beta \to [g_i, \beta] = g_i^{-1}g_i^\beta$ is a homomorphism of B into N with kernel $C_i = C_B(g_i)$. Therefore B/C_i is finitely generated. But $C_1 \cap \cdots \cap C_n = 1$, so B, and hence A, is finitely generated.

Next let A be a radical group of automorphisms of the polycyclic-by-finite group G. We will prove that A is soluble. Let N and B be defined as in the last paragraph and observe that $B \lhd A$. Now a finite radical group is soluble and by the Corollary to Theorem 3.23 a radical linear group is soluble. Therefore, with the aid of an induction on the length of the series in G, we can conclude that A/B is soluble. Now $[G, B, B] = 1$ by definition of B, and by the Three Subgroup Lemma

$$[G, B'] \leq [G, B, B]^G = 1.$$

Hence $B' = 1$ and A is soluble.

To complete the proof it is more than enough to prove that *a soluble group whose subnormal abelian subgroups are finitely generated is polycyclic*. Let H be such a group. If H is abelian, the assertion is obvious, so we can suppose that H is non-abelian; let N be the least non-trivial term of the derived series of H. Then N is finitely generated since it is abelian. Consequently, it is enough to prove that each subnormal abelian subgroup A/N of H/N is finitely generated; for then we can use induction on the derived length of H. Let $C = C_A(N)$. Then $N \leq C$ and A/C is essentially an abelian group of automorphisms of the finitely generated abelian group N. By the first part of the proof, A/C is finitely generated. Now $[C', C] \leqq [N, C] = 1$, so C is nilpotent. Let M be a maximal normal abelian subgroup of C, so that $M \geqq \zeta(C) \geqq N$ and C/M is abelian. Then $M \lhd C \lhd A$ *sn* H, so M *sn* H; hence M is finitely generated. Now M coincides with its centralizer in C, by Lemma 2.19.1. Therefore C/M is finitely generated, by the first part of the proof. It follows that A, and hence A/N, is finitely generated. □

Recently L. Auslander [2] has shown that *the automorphism group of a polycyclic group is finitely presented*; see also Auslander and Baumslag [1].

Groups of Automorphisms of Černikov Groups

It is natural to ask whether there are analogues of these theorems for Černikov groups, for, in the correspondence between Max and Min, polycyclic-by-finite groups correspond to Černikov groups. Here we are immediately faced with an obstacle: there are automorphisms of a group of type p^∞ which have infinite order, so it is not true that a soluble group of automorphisms of a Černikov group is a Černikov group. To avoid this difficulty we restrict attention to periodic groups of automorphisms. In this case the basic lemma is due to Baer.

Lemma 3.28 (Baer [23], p 525). Let G be the direct product of a finite number of groups of type p^∞ and let α be a non-trivial automorphism of G which fixes every element of order p and, when $p = 2$, every element of order 4. Then α has infinite order.

Proof. Suppose that α has finite order; then without loss of generality we can assume that the order of α is a prime q. The group $[G, \alpha] = G^{\alpha-1}$ is an endomorphic image of G, so it is also a direct product of finitely many p^∞-groups. If $G^{\alpha-1} \neq G$, then $G^{\alpha-1}$, being a direct factor of G, is the direct product of fewer groups of type p^∞ than G; also $G^{\alpha-1}$ is α-admissible, so by induction on the number of p^∞-direct factors, either α induces an automorphism of infinite order in $G^{\alpha-1}$, and so in G, or α is trivial on $G^{\alpha-1}$ and $[G, \alpha, \alpha] = 1$. In the latter case

$$[G, \alpha] = [G, \alpha]^q = [G, \alpha^q] = 1$$

and $\alpha = 1$, which is impossible.

We can therefore assume that $G^{\alpha-1} = G$. Let θ denote the endomorphism $1 + \alpha + \alpha^2 + \cdots + \alpha^{q-1}$. Then

$$G^\theta = G^{(\alpha-1)\theta} = G^{\alpha^q-1} = 1$$

and $\theta = 0$. Now let g be an element of least order in G such that $g^\alpha \neq g$. Then $[g, \alpha]^p = [g^p, \alpha] = 1$ by minimality of the order of g, and $g_1 = [g, \alpha]$ has order p. Hence $g_1^\alpha = g_1$ and from $g^\alpha = gg_1$ we infer by induction on i

that $g^{\alpha i} = gg_1^i$. Consequently

$$1 = g^{\theta} = g(gg_1)(gg_1^2) \cdots (gg_1^{q-1}) = g^q g_1^{\binom{q}{2}}. \tag{2}$$

The order of g exceeds p, so (2) shows that $q = p$. If p is odd, p divides $\binom{p}{2}$ and (2) becomes $g^p = 1$. Hence $p = 2$ and from (2) we obtain $g^2 = g_1$, which implies that g has order 4; this is impossible by hypothesis. □

Corollary (Černikov [21]). Let $A = GL(n, R_p)$ be the group of all $n \times n$ matrices with coefficients in R_p (the ring of p-adic integers) and determinant equal to a unit of R_p. Then periodic subgroups of A are finite.

Proof. Let G be the direct product of n groups of type p^∞; then $\operatorname{Aut} G \simeq A$. Let P be a periodic subgroup of $\operatorname{Aut} G$. The elements of order p^2 in G generate a characteristic finite subgroup F and the automorphisms in P that fix every element of F form a subgroup N with finite index in P. By the lemma, N is torsion-free, so $N = 1$ and P is finite. □

We come now to the analogue of Theorem 3.27.

Theorem 3.29 (Baer [23], p. 530, Polovickiĭ [1]). Let A be a periodic group of automorphisms of a Černikov group G. Then A is a Černikov group.

Proof. G has a characteristic subgroup N of finite index such that N is radicable and abelian and satisfies Min. Only finitely many primary components of N are non-trivial and each is a direct product of a finite number of quasicyclic groups. N is A-admissible and, by the Corollary to Lemma 3.28, A induces in N a finite group of automorphisms. Let B denote the subgroup of elements of A that induce trivial automorphisms in both N and G/N. Then $B \lhd A$ and A/B is finite. Now choose a transversal to N in G, say $\{t_1, \ldots, t_n\}$. The map $\beta \to [t_i, \beta]$, $(\beta \in B)$, is a homomorphism of B into N with kernel $C_i = C_B(t_i)$, so B/C_i is an abelian group satisfying Min. An element which lies in each C_i fixes each t_i and also fixes every element of N; hence it is the trivial automorphism of G and $C_1 \cap \cdots \cap C_n = 1$. Therefore B is an abelian group satisfying Min and A is a Černikov group. □

Conversely, Schlette ([1], p. 403) has shown that *a periodic group whose periodic groups of automorphisms are Černikov is itself a Černikov group.*

We shall obtain more precise information about periodic groups of automorphisms of Černikov groups, namely that the subgroup of inner automorphisms is always of finite index. The proof requires a very simple result.

Lemma 3.29.1. Let R be a radicable abelian subgroup and let F be a finite subgroup of a group. If $R^F = R$, then $R = [R, F]\, C_R(F)$.

Proof. Let $x \in R$: since R is abelian, the element

$$y = \prod_{f \in F} x^f$$

belongs to $C_R(F)$. Now, if F has order n,

$$x^{-n} y = \prod_{f \in F} [x, f] \in [R, F],$$

which shows that $x^n \in [R, F]\, C_R(F)$. But $R = R^n$, so the result follows. □

Theorem 3.29.2 (Baer [23], p. 530: see also Černikov [34]). Let G be a Černikov group and let A be a periodic group of automorphisms of G. Then $A/A \cap \operatorname{Inn} G$ is finite.

Proof. Let H be the holomorph of G by A. By Theorem 3.29 the group A is Černikov; hence H is a Černikov group by the Corollary to Theorem 3.12. Let the finite residuals of G, A and H be R, S and T respectively. Then $R \leqq T$ and $S \leqq T$. It is clear that we can find a finite subgroup F such that $G = RF$. By Lemma 3.29.1 we have $T = [T, F]\, C_T(F)$. Now $[T, F] \leqq G$ and $C_T(F) = C_T(G)$ because $G = RF$ and $[R, T] = 1$. Therefore $T \leqq GC_T(G)$ and $S \leqq GC_T(G)$. Thus $S \leqq A \cap \operatorname{Inn} G$; finally, A/S is finite, so $A/A \cap \operatorname{Inn} G$ is finite. □

Corollary. If the centre of G has finite index and G is Černikov, a periodic group of automorphisms of G is finite. In particular this is the case if G is a nilpotent group satisfying Min.

(The second part of the corollary follows from Theorem 3.14.)

For comparison with the Corollary to Theorem 3.28 we mention that *a periodic group of matrices over the field of rational numbers Q is finite.* This may be proved by first showing that a periodic irreducible abelian subgroup of $GL(n, Q)$ is finite (see Lemma 5.29.1) and using the classical theorem of Schur that a periodic group of matrices over the field of complex numbers has a normal abelian subgroup of finite index (see Curtis and Reiner [1], Theorem 36.14). An easy application of this result shows that *a periodic group of automorphisms of a polycyclic-by-finite group is finite*; for this and similar results see Smirnov [4] (Theorem 5).

3.3 The Maximal and Minimal Conditions on Abelian Subgroups

It is clear that Max-*ab*, the maximal condition on abelian subgroups, is equivalent to the property: every abelian subgroup satisfies Max, i.e. is finitely generated. Similarly, Min-*ab*, the minimal condition on abelian subgroups, is equivalent to every abelian subgroup satisfying Min. It is to be expected that Max-*ab* and Min-*ab* are much weaker than Max and Min; certainly every free group satisfies Max-*ab*—for all its abelian subgroups, being free, are cyclic—but does not satisfy Max unless it has rank 1; it seems likely that Min-*ab* is weaker than Min, although, apparently, no example has yet been given.

However in the context of a suitable form of solubility, when abelian subgroups are more numerous, Max-*ab* and Min-*ab* can be expected to play a more decisive part. For example, in [7] (Theorem 8) Mal'cev showed that a soluble group satisfying Max-*ab* is polycyclic and O. J. Schmidt ([5], Theorem 9) showed that a soluble or even hyperabelian group with Min-*ab* is a Černikov group. Thus for soluble groups Max and Max-*ab* are the same property, as are Min and Min-*ab*. These results have inspired a whole series of group theorists to study problems involving groups whose abelian subgroups are subject to finiteness conditions.

Here we will prove the following generalizations of the Mal'cev-Schmidt theorems.

Theorem 3.31. A radical group whose Hirsch-Plotkin radical satisfies Max-*ab* is polycyclic.

Theorem 3.32. A radical group whose Hirsch-Plotkin radical satisfies Min-*ab* is a soluble Černikov group.

Thus we obtain the result of Plotkin [10] that a radical group which satisfies Max-*ab* (Min-*ab*) is a polycyclic (Černikov) group. In the case of hyperabelian groups satisfying Max-*ab* this was proved by Baer ([24], Satz B). We remark that Černikov has shown that a locally soluble group with Min-*ab* is a Černikov group—see Theorem 3.45 below. This is not a special case of Theorem 3.32.

The proofs of Theorems 3.31 and 3.32 depend in an essential way on results in Section 3.2.

Proof of Theorem 3.31

(i) If a group G satisfies Max-ab and if N is a normal polycyclic-by-finite subgroup of G, then G/N satisfies Max-ab.

Let A/N be an abelian subgroup of G/N and let $C = C_A(N)$. It is easy to see that A—and hence A/C—is soluble-by-finite. Therefore A/C is polycyclic-by-finite, by Theorem 3.27. Now $[C', C] \leqq [N, C] = 1$, so C is nilpotent. Let M be a maximal normal abelian subgroup of C. Then M is finitely generated and coincides with its centralizer in C. Hence C/M is polycyclic and consequently A is polycyclic-by-finite, which shows that A/N is finitely generated.

(ii) If G is a locally nilpotent group satisfying Max-ab, then G is polycyclic and therefore nilpotent.

It is enough to establish the existence of a maximal finitely generated subgroup M of G. For then if $g \in G$, the subgroup $\langle g, M\rangle$ finitely generated and hence $g \in M$; thus $G = M$, so G is finitely generated and nilpotent, and therefore polycyclic. If G has no such subgroup, there is a countably infinite, ascending chain of finitely generated nilpotent subgroups of G,

$$N_1 < N_2 < \cdots.$$

The union of this chain, N, obviously cannot be finitely generated.

Let Z_i denote the centre of N_i. The group generated by all the Z_i is abelian and therefore finitely generated, so there is a integer i such that $Z_j \leqq \langle Z_1, \ldots, Z_i \rangle$ for all $j \geqq i$. If $i \leqq j \leqq j_1$, then $Z_{j_1} \leqq N_i \leqq N_j \leqq N_{j_1}$, so $Z_{j_1} \leqq Z_j$. Therefore

$$Z_i \geqq Z_{i+1} \geqq \cdots.$$

Suppose first that N is not torsion-free; then for all sufficiently large j the group N_j is not torsion-free and hence Z_j is not torsion-free; for Z_j must intersect the torsion-subgroup of N_j non-trivially by Lemma 2.16. Let T_j be the torsion-subgroup of Z_j: then for some large integer j

$$T_j \geqq T_{j+1} \geqq T_{j+2} \geqq \cdots > 1.$$

However T_j, being a finitely generated periodic abelian group, is finite. It follows that there is an integer k such that $T_k = T_{k+1} =$ etc. and $\zeta(N) \geqq T_k > 1$. Now suppose that N is torsion-free. By Theorem 2.25, the group N_{p+1}/Z_{p+1} is torsion-free and so Z_p/Z_{p+1} is torsion-free for all $p \geqq i$. Thus Z_p/Z_{p+1} is a free abelian group of finite rank, and if $Z_p \neq Z_{p+1}$, then Z_{p+1} has smaller rank than Z_p. Now the ranks of the Z_p are positive and cannot decrease indefinitely. Hence there is a p for which

$$Z_p = Z_{p+1} = \text{etc}$$

and again $\zeta(N) \geqq Z_p > 1$.

Thus in either case $\zeta(N)$ is non-trivial: it is also finitely generated and therefore polycyclic. The first part of the proof shows that $N/\zeta(N)$ also satisfies Max-*ab*. We form the chain $\{N_i\zeta(N)/\zeta(N): i = 1, 2, \ldots\}$, omitting any repetitions, and note that the argument just used implies that $N/\zeta(N)$ has non-trivial centre. Indeed in this way we can show that

$$\zeta_1(N) < \zeta_2(N) < \cdots < \zeta_i(N) < \zeta_{i+1}(N) < \cdots$$

where i is finite. Set $L = \zeta_\omega(N)$, a hypercentral group, and let A be a maximal normal abelian subgroup of L. Then $A = C_L(A)$, while A is finitely generated and L/A is a hypercentral group of automorphisms of A. Obviously hypercentral groups are radical, so Theorem 3.27 may be applied to show that L/A, and hence L, is finitely generated. But this implies that $L = \zeta_i(N)$ for some finite i, which is a contradiction.

(iii) The general case.

Let G be a radical group and let R, the Hirsch-Plotkin radical of G, satisfy Max-*ab*. By *(ii)* R is polycyclic. In addition $C_G(R) \leqq R$ by Lemma 2.32. Now $G/C_G(R)$ is a radical group of automorphisms of R, so it must be polycyclic by Theorem 3.27. Thus G/R, and hence G, must be polycyclic. □

Proof of Theorem 3.32

(i) If a group G satisfies Min-ab and if N is a normal Černikov subgroup of G, then G/N satisfies Min-ab.
The proof is similar to the corresponding part of the proof of Theorem 3.31.

(ii) If G is a locally nilpotent group satisfying Min-ab, then G is a hypercentral Černikov group.
The main difficulty resides in proving that G is hypercentral: suppose that this is not the case. Then, by the remark following Theorem 2.19, we may suppose that G is countable. Since G is locally nilpotent, there is a countably infinite, ascending chain of finitely generated (and hence finite) nilpotent subgroups

$$N_1 < N_2 < \cdots$$

with its union equal to G. Let $Z_i = \zeta_i(N)$ and set $Z = \langle Z_1, Z_2, \ldots \rangle$. Then Z is abelian, so it satisfies Min and the elements of square-free order generate a finite subgroup P. Clearly $N_i \neq 1$ implies that $P \cap Z_i \neq 1$. Now since P is finite, $P \leqq N_i$ for some integer i and if $i \leqq j \leqq j_1$, then $P \cap Z_{j_1} \leqq N_i \leqq N_j \leqq N_{j_1}$, so that $P \cap Z_{j_1} \leqq P \cap Z_j$. Consequently

$$P \cap Z_i \geqq P \cap Z_{i+1} \geqq \cdots > 1.$$

Since P is finite, there is an integer j such that $P \cap Z_j = P \cap Z_{j+1} = \text{etc.}$ Hence $\zeta(G) \geqq P \cap Z_j > 1$. Let α be any ordinal and suppose that $\zeta_\alpha(G) \neq G$. Now $L = \zeta_\alpha(G)$ is a hypercentral group satisfying Min-*ab*; since maximal normal abelian subgroups of L are self-centralizing, Theorem 3.29 can be applied to show that L is a Černikov group. G/L inherits the property Min-*ab* from G by *(i)* and G/L is the union of the chain of nilpotent subgroups $\{N_iL/L : i = 1, 2, \ldots\}$, yet it not itself hypercentral. Consequently, our argument shows that G/L has non-trivial centre and $L = \zeta_\alpha(G) < \zeta_{\alpha+1}(G)$. It follows that G is hypercentral.

By this contradiction our original locally nilpotent group G is hypercentral and by an argument used in the previous paragraph the property Min-*ab* implies that G is a Černikov group.

(iii) The general case.
Let G be a radical group and assume that its Hirsch-Plotkin radical R satisfies Min-*ab*. Unfortunately we cannot simply imitate the proof of Theorem 3.31 at this point, because it is not clear that G is periodic and our theorem on periodic automorphism groups of a Černikov group is not immediately applicable. We proceed in a slightly different manner.

R is a Černikov group by *(ii)*: let F denote its finite residual. Let C be the subgroup of elements of G that centralize R/F and commute with each element of prime order in F. Then $C \lhd G$ and, since F is an abelian group satisfying Min, G/C is finite. It is enough to prove that $C \leqq R$, so let us assume that this is not the case. Since G is radical, i.e. hyper-(locally nilpotent), there is a normal subgroup L of G such that $R < L \leqq CR$ and L/R is locally nilpotent. $L = L \cap (CR) = (L \cap C) R$ and $L \cap C \lhd G$; hence $L \cap C \nleqq R$. Thus $L \cap C$ cannot be locally nilpotent, and it therefore possesses a finitely generated non-nilpotent subgroup, say X. Let a be an element of F with order p^i where p is a prime and $i > 0$, and let $x \in X$. Then $[a, x]^{p^{i-1}} = [a^{p^{i-1}}, x] = 1$ since X centralizes all elements of prime order in F; therefore $[a, x]$ has order dividing p^{i-1}. Now let F_i be the subgroup of F generated by all elements of order $\leq p^i$ where p is any prime; then $F_0 = 1$ and $[F_i, x] \leqq F_{i-1}$ if $i > 0$. It follows that $X \cap R$ lies in the hypercentre of X and this, together with the nilpotence of XR/R, shows that X is actually hypercentral. But hypercentral groups are locally nilpotent (see p. 50) and X is finitely generated, so X is nilpotent, which is a contradiction. ▯

In the remainder of this section we consider some other situations when the maximal or minimal condition on abelian subgroups implies the corresponding condition on all subgroups without a hypothesis of solubility.

Theorem 3.33 (Baer [23], p. 530). A group which satisfies Min-*ab* is a Černikov group if and only if it has a maximal abelian subgroup with only finitely many conjugates.

Proof. Let G be a Černikov group; the finite residual of G is abelian, so it is contained in a maximal abelian subgroup which has finite index and therefore only finitely many conjugates in G.

Conversely let G satisfy Min-*ab* and let A be a maximal abelian subgroup which has only a finite number of conjugates in G. Let $N = N_G(A)$ and $C = C_G(A)$. Then $|G: N|$ is finite by hypothesis and N/C is also finite since it is a periodic group of automorphisms of A. Hence $|G: C|$ is finite: but obviously $C = A$ since A is maximal abelian. Therefore G is a Černikov group. ▯

The next result, although of a similar character, does not depend on the theory of periodic automorphism groups of Černikov groups but reflects the special nature of groups of finitary permutations.

Theorem 3.34. Let G be a group of finitary permutations of a set X (that is, each element of G fixes all but a finite number of the elements of X) and assume that G has a maximal abelian subgroup which satisfies Min. Then G is finite.

Corollary. A group of finitary permutations which satisfies Min-*ab* is finite.

This generalizes a result of Ado [2] according to which a subgroup of the restricted countable symmetric group which satisfies Min is finite: see also Scott [5], p. 304.

We need the following lemma for the proof of Theorem 3.34.

Lemma 3.35 (Scott [5], 11.2.7). Let G be an infinite group of finitary permutations of a set X and let Y be a finite subset of X. Then $F_G(Y)$, the subgroup of elements of G fixing every element of Y, is infinite.

Proof. We will show first that $F_G(x)$ is infinite for each $x \in X$. Suppose that this is not the case and $F_G(x)$ is finite; then, since G is infinite, x belongs to an infinite G-orbit U. Let g be a given element of G such that $xg \neq x$ and let $S(g)$ be the set of all elements of X which are moved by g. Let $y \in X$ and $h \in G$; then $yg \neq y$ is equivalent to $(yh)\, g^h \neq yh$. It follows that

$$S(g^h) = (S(g))\, h \tag{3}$$

and this implies that $S(g^h)$ has the same finite number of elements as $S(g)$, say n. Now $x \in S(g)$, so (3) also shows that

$$U \subseteq \bigcup_{h \in G} S(g^h).$$

Since U is infinite, g has infinitely many distinct conjugates in G. Let $x_1, \ldots, x_{n+1}$ be distinct elements of U: each g^h moves exactly n elements of X and hence must fix some x_i, that is $g^h \in F_G(x_i)$. Therefore the conjugates of g lie in the union of the $F_G(x_i)$. But $F_G(x_i)$ is conjugate to $F_G(x)$ and hence is finite, because x_i and x lie in the same G-orbit; this implies that g has only a finite number of conjugates in G, which is a contradiction.

Now let $Y = \{y_1, \ldots, y_m\}$ and let S_i denote $F_G(\{y_1, \ldots, y_i\})$; then $S_i = F_{S_{i-1}}(y_i)$, if $i > 1$. By the result of the preceding paragraph S_1 is infinite and if S_{i-1} is infinite, so is S_i. Hence $S_m = F_G(Y)$ is infinite. $\square$

Proof of Theorem 3.34. Let M be a maximal abelian subgroup of G and let M satisfy Min. Then M *is even finite*: to prove this it will be enough to show that G has no subgroups of type p^∞ for any prime p. Let a be a non-trivial element of G with order a power of p and suppose that $a = b^{p^i}$ where $b \in G$ and $i > 0$. Write a and b as products of disjoint cycles of length > 1 and let $n(a)$ and $n(b)$ be the respective numbers of cycles in these products. The length of each cycle is a power of p and the p^ith power of a cycle of length p^m is a product of p^i cycles of length p^{m-i} if $m > i$. Hence, by enumeration of the cycles of length > 1, we obtain

$$n(a) = p^i(n(b) - n_1(b) - \cdots - n_i(b)),$$

where $n_j(b)$ is the number of cycles of length p^j in the decomposition of b. Hence p^i divides $n(a)$ and a cannot be a p^ith power for arbitrary i. Thus G has no subgroup of type p^∞.

Let Y denote the set of elements of X which are moved by at least one element of M. Since M is finite, so is Y. Suppose that G is infinite; then by Lemma 3.35 the subgroup $S = F_G(Y)$ is infinite. Let $a \in M$, $s \in S$ and $x \in X$: if $x \in Y$, then $xas = xs^{a^{-1}}a = xa = xsa$ since $S^M = S$; if $x \notin Y$, then $xas = xs = xsa$ since $xs \notin Y$. Hence $as = sa$ and $S \leq C_G(M)$. But $C_G(M) = M$ because M is a maximal abelian subgroup. Thus $S \leq M$, yet S is infinite and M is finite. □

If N is a normal subgroup of a group G, we write

$$\mathrm{Aut}_G(N)$$

to be the group of all automorphisms of N which arise through transformation by elements of G. Naturally

$$\mathrm{Aut}_G(N) \simeq G/C_G(N).$$

Theorem 3.36 (cf. Baer [41], Zusatz 2). Let $\mathfrak{X}$ be a class of groups such that $\mathfrak{X} = \boldsymbol{H}\mathfrak{X} = \boldsymbol{R}_0\mathfrak{X}$. Let G be a group such that if H is a non-trivial homomorphic image of G, there is a non-trivial normal subgroup N of H such that $\mathrm{Aut}_H(N')$ belongs to $\mathfrak{X}$. If G satisfies Max-*ab* (Min-*ab*), then G is polycyclic-by-$\mathfrak{X}$ (Černikov-by-$\mathfrak{X}$).

Proof. Suppose that G is a group with Max-*ab*—for groups with Min-*ab* the proof is nearly identical. Let R be the $\mathfrak{S}$-radical of G, i.e. the product of all the normal soluble subgroups of G. Clearly R is hyperabelian, so in particular it is radical and, in view of Theorem 3.31, R is polycyclic. In addition G/R has no non-trivial normal soluble subgroups. Since G/R inherits the properties of G—including Max-*ab*, by the first part of the proof of Theorem 3.31—we may assume that G has no non-trivial normal soluble subgroups. Suppose that $G \notin \mathfrak{X}$.

By hypothesis there is a non-trivial normal subgroup of G whose derived subgroup N is such that $\mathrm{Aut}_G N \in \mathfrak{X}$; also $N \neq 1$. Let $C = C_G(N)$; then $G/C \in \mathfrak{X}$, so we can be sure that $C \neq 1$. Since $C \cap N$ is abelian and normal in G, we must have $C \cap N = 1$. By Zorn's Lemma there is a normal subgroup M of G which is maximal subject to $C \cap M = 1$. Now $M \neq G$ since $C \neq 1$; hence there is a non-trivial normal subgroup L/M of G/M such that

$$\mathrm{Aut}_{G/M}(L/M)' = \mathrm{Aut}_G(L/M)' \in \mathfrak{X}.$$

The maximality of M shows that $C \cap L \neq 1$. If L/M were abelian, $C \cap L$ would be abelian because $C \cap L \simeq (C \cap L)\,M/M \leq L/M$: this is impossible since $C \cap L \lhd G$. Therefore L/M is not abelian and $M < L'M$; this shows that

$$D = C \cap (L'M) \neq 1.$$

Now $D \cap M = 1$, so the natural homomorphism of D onto DM/M is an isomorphism and, obviously, it is even a G-operator isomorphism. Therefore $\mathrm{Aut}_G D \simeq \mathrm{Aut}_G(DM/M)$, which is a homomorphic image of $\mathrm{Aut}_G(L/M)'$ and hence belongs to $\mathfrak{X}$ by $\boldsymbol{H}$-closure.

Let $N_1 = N$ and $N_2 = D$: these are non-trivial normal subgroups of G with trivial intersection, so their product is direct. In addition they have the properties $\mathrm{Aut}_G N_1 \in \mathfrak{X}$ and $\mathrm{Aut}_G N_2 \in \mathfrak{X}$. Obviously $C_G(N_1 \times N_2) = C_G(N_1) \cap C_G(N_2)$, so

$$\mathrm{Aut}_G(N_1 \times N_2) \in \boldsymbol{R}_0\mathfrak{X} = \mathfrak{X}.$$

$N_1 \times N_2$ has all the properties of the original normal subgroup $N_1 = N$, so the same argument can be applied to $N_1 \times N_2$. By repeated application of this procedure we can generate a sequence of non-trivial normal subgroups $N_1, N_2, \ldots$ such that for each integer i the product of $N_1, \ldots, N_i$ is direct and $\mathrm{Aut}_G(N_1 \times \cdots \times N_i) \in \mathfrak{X}$. However this contradicts Max-*ab* since if $1 \neq x_i \in N_i$, the x_i generate an abelian subgroup which is not finitely generated. □

Corollary (cf. Baer [41], p. 175). A group G satisfies Max (Min) if and only if it satisfies Max-*ab* (Min-*ab*) and each non-trivial homomorphic image H of G contains a non-trivial normal subgroup N such that $\mathrm{Aut}_H(N')$ satisfies Max (Min).

To prove the Corollary let $\mathfrak{X}$ be the class of groups which satisfy Max (Min) and note that we have $\boldsymbol{H}$ and $\boldsymbol{R}_0$-closure (see Lemma 1.48).

Some related theorems are in the paper of Baer [41] and a special case of the corollary is in Simon [3].

3.4 Locally Finite Groups and Schmidt's Problem. The Hall-Kulatilaka-Kargapolov Theorem

An infinite group all of whose proper subgroups are finite is called *quasifinite*; clearly groups of type p^∞, for all primes p, are quasifinite. *Whether every quasifinite group is of type p^∞ for some prime p is unknown*, and this is usually referred to as *Schmidt's Problem* (after O. J. Schmidt).

More generally one might ask *if there are any infinite groups having all their abelian subgroups finite*, i.e. satisfying Max-*ab* and Min-*ab*. However, as has already been mentioned in Chapter 1 (p. 35), the Burnside groups of high prime exponent have this property; these groups have proper infinite subgroups.

Not surprisingly the 2-groups, in contrast to the p-groups for $p \geqq 4381$, are well-behaved.

Theorem 3.41 (Held [4], Kegel [5], Strunkov [3], Theorem 8). Every infinite 2-group contains an infinite abelian subgroup. The only quasi-finite 2-groups are the groups of type 2^∞.

Proof. Let G be an infinite 2-group. By the Corollary to Theorem 3.15, we can find an infinite ascending chain of finite subgroups of G, say $F_1 < F_2 < \cdots$. Let U be the union of the chain and notice that U is an infinite locally finite 2-group, so it is locally nilpotent. Suppose that G has no infinite abelian subgroups; then U satisfies Min-*ab* and Theorem 3.32 shows that U is a Černikov group. Since U cannot have a subgroup of type 2^∞, it must be finite and this is a contradiction. □

A much deeper theorem, discovered independently by Hall and Kulatilika [1] and Kargapolov [11], asserts that every infinite locally finite group contains an infinite abelian subgroup.

We shall prove first an interesting theorem of Šunkov.

Theorem 3.42 (Šunkov [5], see also Kegel [8], p. 42, Theorem 4.3). Let G be an infinite periodic group containing an involution. Then either the centre of G contains an involution or G has a proper infinite subgroup with non-trivial centre.

Proof. We can assume that every involution in G has a finite centralizer in G since otherwise the truth of the theorem is evident. Let i be an involution in G: a non-trivial element of G which is transformed into its inverse by i will be called *i-involuted*. The first point to establish is that there are infinitely many i-involuted elements in G. Since $C_G(i)$ is finite, $|G: C_G(i)|$ is infinite and therefore G contains infinitely many involutions, for example the conjugates of i. Hence there exist infinitely many distinct cosets $C_G(i)a$, different from $C_G(i)$, where a is an involution. Let $z = aa^i$; then $z^i = a^i a = z^{-1}$ and $z \neq 1$ since $a^i \neq a$. Also $aa^i = bb^i$ and $a^2 = 1 = b^2$ imply that $C_G(i)a = C_G(i)b$; therefore the assertion is proved. Indeed there exist infinitely many i-involuted elements in G with *odd order*. For otherwise there is an infinite sequence of i-involuted elements of even order, say $x_1, x_2, \ldots$, and for each j there is an integer m_j such that $x_j^{m_j}$ is an i-involuted involution, which, of course, means that i centralizes each $x_j^{m_j}$ and that $\{x_1^{m_1}, x_2^{m_2}, \ldots\}$ is a finite set of involutions; consequently there is an involution in this set which is centralized by infinitely many of the elements x_j, and this is contrary to our hypothesis.

Let S be the set of all i-involuted elements of odd order and let a be a fixed element of S. Write $k = ia$ and note that $k^2 = i^2 a^i a = a^{-1} a = 1$ and $a \neq i$; hence k is an involution. Let b be any element of S and set

$u = ik^b$; then

$$u^i = k^b i = (ik^b)^{-1} = u^{-1}.$$

$C_G(k)$ is finite, so, by allowing b to take values in S, we obtain an infinity of distinct elements u. Moreover, those b in S for which u has odd order form an infinite subset T of S; otherwise, just as before, we would be able to find an involution with infinite centralizer.

Let b be any element of T and as before write $u = ik^b$. Since $u^i = u^{-1}$ and u has odd order, $\langle u, i \rangle$ is a finite dihedral group with Sylow 2-subgroups of order 2. Now $k^b \in \langle u, i \rangle$, so Sylow's Theorem shows that there is an element u_1 in $\langle u \rangle$ such that

$$i = (k^b)^{u_1}.$$

The same reasoning applied to the group $\langle a, i \rangle$ yields

$$i = k^{a_1}$$

for some $a_1 \in \langle a \rangle$. Combining these equations we obtain $k^{bu_1 a_1^{-1}} = k$; therefore $h = bu_1 a_1^{-1} \in C_G(k)$ and $u_1 = b^{-1} h a_1$. Now let $g = b^{-1} h = u_1 a_1^{-1}$, so that $u_1 = g a_1$ and $u_1^i = g^i a_1^i$. But u_1 and a_1 are powers of u and a respectively; therefore $u_1^i = u_1^{-1}$ and $a_1^i = a_1^{-1}$. It follows that

$$u_1^{-1} = a_1^{-1} g^{-1} = u_1^i = g^i a_1^{-1}$$

and therefore $g^{ia_1^{-1}} = g^{-1}$. Let $j = ia_1^{-1}$; then j is an involution and $g^j = g^{-1}$. Finally, by writing $\bar{h} = h^{-1}$ we obtain $(g^{-1})^j = g$ and $(\bar{h}b)^j = b^{-1}\bar{h}^{-1} = (\bar{h}b)^{-1}$.

What we have shown is that for each $b \in T$ there is an $\bar{h} \in C_G(k)$ such that $\bar{h}b$ is j-involuted where j is an involution of the form ia_1^{-1}, $a_1 \in \langle a \rangle$. Notice that i and a—and therefore k, a_1 and j—are fixed throughout this calculation and do not depend on b. However $C_G(k)$ is finite, so there is an infinite sequence of distinct elements of T, say $b_1, b_2, \ldots$, such that for some $c \in C_G(k)$ the elements $cb_1, cb_2, \ldots$ are j-involuted. If $m > 1$, then $(cb_1)^j = b_1^{-1}c^{-1}$ and $(cb_m)^j = b_m^{-1}c^{-1}$, and these imply that $(b_m^{-1}b_1)^j = (b_m b_1^{-1})^{c^{-1}}$. Now b_1 and b_m are i-involuted and $j = ia_1^{-1}$, so $(b_m^{-1}b_1)^j = (b_m b_1^{-1})^{a_1^{-1}}$. Comparison of the two expressions for $(b_m^{-1}b_1)^j$ yields $(b_m b_1^{-1})^{a_1^{-1}c} = b_m b_1^{-1}$. Since the elements $b_m b_1^{-1}$, $m = 2, 3, \ldots$, are all distinct, $d = a_1^{-1}c$ has infinite centralizer. Suppose that $C_G(d) = G$. Then $d^k = d$ and, since $c \in C_G(k)$, it follows that $a_1^k = a_1$. Now

$$a_1^k = a_1^{ia} = a_1^{-1},$$

because $a_1 \in \langle a \rangle$; hence $a_1^2 = 1$ and therefore $a_1 = 1$ since a has odd order. Consequently $i = k^{a_1} = k = ia$ and $a = 1$, which is impossible. Therefore $C_G(d)$ is a proper infinite subgroup of G and $d \neq 1$. $\square$

Corollary. A quasifinite group which contains an involution has non-trivial centre.

Results related to Theorem 3.42 are in the paper [9] of Šunkov.

Theorem 3.43 (Hall and Kulatilaka [1], Kargapolov [11]). An infinite locally finite group contains an infinite abelian subgroup.

The proof of this theorem is definitely not elementary, as it involves the Feit-Thompson Theorem on the solubility of groups of odd order; it also requires a theorem of Burnside on Frobenius groups and the theorem of Mal'cev that simple locally soluble groups are cyclic.*

Proof of Theorem 3.43

(i) It is sufficient to show that in each infinite locally finite group there is a non-trivial element with infinite centralizer.
For assume that this has been proved and let G be an infinite locally finite group. Let A_1 be a finite abelian subgroup with infinite centralizer in G—for example $A_1 = 1$ will do. Let $C_1 = C_G(A_1)$; then $A_1 \lhd C_1$ and C_1/A_1 is infinite, so there is an element x in $C_1 \backslash A_1$ such that the group

$$C_{C_1/A_1}(xA_1) = D/A_1$$

is infinite. Set $A_2 = \langle x, A_1 \rangle$; then $A_1 < A_2$ and A_2 is abelian, so we can assume that it is finite. Let $C_2 = C_G(A_2)$; then $C_2 \leqq D$ and hence $C_2 \leqq C_D(x)$. Also, since $D \leqq C_1$, we have $C_D(x) \leqq C_2$ and therefore $C_2 = C_D(x)$. Consequently

$$|D: C_D(x)| = |D: C_2| \leqq |\mathrm{Aut}\, A_2| < \infty,$$

since by definition D normalizes A_2. This with the fact that D is infinite shows that $C_2 = C_D(x)$ is infinite. In this way we can find an infinite ascending chain of finite abelian subgroups $A_1 < A_2 < \cdots$ such that each $C_G(A_i)$ is infinite. The union of this chain is an infinite abelian subgroup of G.

Accordingly, let G be an infinite locally finite group such that $C_G(x)$ is finite for each $1 \neq x \in G$. We look for a contradiction.

(ii) If F is a non-trivial finite subgroup of G, then $N_G(F)$ is finite.
Let $1 \neq a \in F$: then $C_G(F) \leqq C_G(a)$, which is finite, so $C_G(F)$ is finite. $N_G(F)/C_G(F)$ is clearly finite and it follows that $N_G(F)$ is finite.

(iii) There is a finite subgroup F such that $C_G(F) = 1$.
Let $1 \neq a \in G$ and let $1, b_1, \ldots, b_n$ be the distinct elements of $C_G(a)$. Each $C_G(b_i)$ is finite, so we can choose an element $c_i \notin C_G(b_i)$ for each i. Let $F = \langle a, b_1, \ldots, b_n, c_1, \ldots, c_n \rangle$, a finite subgroup. Then $C_G(F) \leqq C_G(a)$

* For an elementary proof of Mal'cev's theorem see Corollary 1, Theorem 5.27 and also vol. 2, p. 83.

and yet no element of $C_G(a)$ except 1 centralizes F. Therefore $C_G(F) = 1$.

(iv) For each prime p the Sylow p-subgroups of G are finite and conjugate.

Let P be an infinite Sylow p-subgroup of G. Since P is also locally finite, there is a finite subgroup F of P such that $C_P(F) = 1$. But F is a non-trivial finite p-group; hence $1 < \zeta(F) \leqq C_P(F)$. Therefore P is finite. Two Sylow p-subgroups of G generate a finite group of which they are also Sylow p-subgroups; by Sylow's Theorem they are conjugate in this group.

(v) If $M \lhd G$ and P is a Sylow p-subgroup of M, then $G = N_G(P)\, M$.

Let $x \in G$; now P and P^x are Sylow p-subgroups of M and by (iv) they are conjugate in M, say $P^x = P^y$ for some $y \in M$. Then $xy^{-1} \in N_G(P)$ and $x \in N_G(P)\, M$.

(vi) Every proper homomorphic image of G is finite.

For let $1 \neq M \lhd G$ and let P be a non-trivial Sylow subgroup of M. Then P is finite by *(iv)*, and by *(ii)* $N_G(P)$ is finite. *(v)* shows that $G = N_G(P)\, M$, so G/M is finite.

(vii) G contains no involutions and is locally soluble.

If G contained an involution, it would have a non-trivial element with infinite centralizer, by Šunkov's theorem (3.42). Hence finitely generated subgroups of G have odd order and are soluble by the Feit-Thompson Theorem [1].

Now comes the main step in the proof.

(viii) G is not residually finite.

Suppose that G is residually finite. Let P be a non-trivial Sylow p-subgroup of G and define

$$T = \langle C_G(x) : 1 \neq x \in P \rangle .$$

Then $P \leqq T$ and T is finite since P is finite. From the residual finiteness of G it follows that there is a normal subgroup K with finite index in G, such that $K \cap T = 1$. Clearly $K \cap P = 1$, so K has no elements of order p. Let Q be a non-trivial Sylow q-subgroup of K and let $N = N_G(Q)$. Then $p \neq q$ and $G = NK$ by (*v*). Now $|P|$ divides $|G : K| = |N : N \cap K|$ since $K \cap P = 1$. Hence Sylow p-subgroups of N have the same order as P and so are conjugate to P. Replacing P by a suitable conjugate if necessary, we can assume that $P \leqq N$ and $Q^P = Q$. Suppose that $a^b \in P$ where $1 \neq a \in P$ and $1 \neq b \in Q$. Then $[a, b] \in Q \cap P = 1$ and $a^b = a$; this implies that $b \in T$, by definition of T, and hence that $b \in K \cap T = 1$. From this we deduce that

$$P \cap P^b = 1, \quad \text{if} \quad 1 \neq b \in Q .$$

This means that PQ is a *Frobenius group* and, p being odd, a classical result of Burnside shows that P is cyclic (Scott [5], Theorem 12.6.15).

We have just proved that every Sylow subgroup of G is cyclic. Now a finite group with cyclic Sylow subgroups is metabelian (i.e. of derived length $\leqq 2$), by another well-known result of Burnside (Scott [5], Theorem 12.6.17). Consequently G is locally metabelian and therefore metabelian. Obviously G cannot be abelian, so G/G' is finite by *(vi)*; but G' is abelian, so it is also finite and hence G is finite.

(ix) The final stage.
Let R be the finite residual of G. By *(viii)* $R \neq 1$ and by *(vi)* G/R is finite. Thus R is infinite and must therefore satisfy the initial hypotheses on G. Let $1 \neq N \lhd R$: then R/N is finite and $|G:N|$ is finite, which implies that $N = R$. Therefore R is simple. But R is also locally soluble by *(vii)* and Mal'cev has shown that a simple locally soluble group is cyclic of prime order (Theorem 5.27, Corollary 1). Hence G is finite. □

Strunkov ([3], Theorem 7) has generalized the Hall-Kulatilaka-Kargapolov Theorem, showing that *an infinite group in which every 2-generator subgroup is finite contains an infinite abelian subgroup.*

It may be as well to review the status of Schmidt's Problem in the light of the information now available to us. Suppose that G is a quasifinite group which is not of type p^∞ for any prime p. Clearly G satisfies Min. Also G is finitely generated, since otherwise finitely generated subgroups of G would be proper and hence finite and G would be locally finite, which we know cannot be the case. If $G_1 < G_2 < \cdots$ were an infinite ascending chain of subgroups of G, its union would have to be G, and yet this is impossible since G is finitely generated. Therefore G also satisfies Max. Let N be a maximal normal subgroup of G and set $H = G/N$; then N is finite, so H is infinite and it is obviously also quasifinite and it satisfies Max and Min. Finally H is simple, so it can have no involutions, by Šunkov's Theorem. Thus Schmidt's Problem reduces to the following.

Does there exist an infinite simple group without elements of order 2, with every proper subgroup finite, and satisfying Max as well as Min?

By Strunkov's generalization of Theorem 3.43 this group would also have to be a 2-generator group.

Thus we arrive back at the conjectures that groups which satisfy Max are polycyclic-by-finite and that groups which satisfy Min are Černikov groups. If either of these conjectures were true, a group of the above type could not exist and Schmidt's Problem would be solved. Just how far we are from solving these problems can be seen from the fact that *it is not known if there exist infinite groups in which every proper non-trivial subgroup has order p where p is an odd prime.** Such a group would be a very special example of a quasifinite group with the properties listed above. In this connection see the paper of Newman and

* This question is generally attributed to Tarski.

Wiegold [1] on minimal non-nilpotent* groups and also Zel'manzon [1] on minimal non-cyclic groups; finite minimal non-abelian and non-nilpotent groups have been classified by Miller and Moreno [1] and O. J. Schmidt [2] respectively.

Further Results about Locally Finite Groups

A number of recent articles by Šunkov and an article by Kegel and Wehrfritz have greatly added to our knowledge of locally finite groups subject to various minimal conditions. On the basis of work of Šunkov [3], [4], [8], and of work in the theory of finite groups due to Alperin, Brauer and Gorenstein [1], Bender [1] and Brauer [1], it can be shown that *a locally finite group satisfying Min is a Černikov group*, a result that had long been conjectured.

More generally Šunkov has recently proved that *a locally finite group with Min-ab is a Černikov group* (Šunkov [10]), and Kegel and Wehrfritz [1] have shown that *a locally finite group which satisfies either the minimal condition on centralizers of subgroups or the minimal condition on p-subgroups for every prime p is abelian-by-finite.*

As a general rule the proofs of these results involve deep theorems from finite group theory, and it is reasonable to expect such methods to play an increasing role in the theory of locally finite groups. Further papers which may be consulted are Kegel [4], [6], [7] and Šunkov [6]. The set of lecture notes of Kegel [8] is cited as a general reference for locally finite groups.

Applications of the Hall-Kulatilaka-Kargapolov Theorem

The key to the problem of groups satisfying Min is of course the structure of the minimal non-Černikov groups. Amberg has used the Hall-Kulatilaka-Kargapolov Theorem to prove the following result concerning such groups.

Theorem 3.44 (Amberg [3], Lemma 3). Let G be a minimal non-Černikov group. Then G has at least one maximal subgroup and each maximal subgroup has infinite index and normalizes no proper subgroups of G which it does not contain.

Proof. First we show that G *has* a maximal subgroup. If G is finitely generated, this is obvious, so we can assume that finitely generated subgroups of G are proper and hence finite.** By Zorn's Lemma there is a maximal radicable abelian subgroup of G, say R. Suppose that $R \leq H < G$;

* If $\mathscr{P}$ is a group theoretical property, a *minimal non-$\mathscr{P}$-group* is a group which does not have $\mathscr{P}$ but each of whose proper subgroups has $\mathscr{P}$.

** We do not, of course, want to use the theorem that a locally finite group with Min is a Černikov group. This would show that G is finitely generated.

then H is a Černikov group and R must lie in the finite residual of H. But the latter is a radicable abelian group, so R must be the finite residual of H and consequently $R \lhd H$ and H/R is finite. Suppose that R does not lie in a maximal subgroup of G; then $R \lhd G$ and G/R is locally finite. Let A/R be an abelian subgroup of G/R. A minimal non-Černikov group satisfies Min, so it cannot be soluble, by Theorem 3.12. Hence $A < G$ and consequently A/R is finite. Theorem 3.43 now shows that G/R is finite, which is impossible, for it would make G a Černikov group. Hence R is contained in a maximal subgroup of G.

Now let M be any maximal subgroup of G. Certainly M has infinite index in G, for otherwise G would have a proper normal subgroup of finite index. Suppose that M normalizes a proper subgroup S such that $S \nleqq M$. Then $S \lhd \langle M, S\rangle = MS$ and $G = MS$, so $S \lhd G$. But both M and S are Černikov groups and the latter form a $\boldsymbol{P}$-closed class (by the Corollary to Theorem 3.12), so G is a Černikov group. ▯

Corollary. A group G is a Černikov group if and only if it satisfies Min and each subgroup has no maximal subgroups with infinite index.

Proof. We omit the easy proof that each maximal subgroup of a Černikov has finite index. To prove the sufficiency of the condition we merely choose a smallest counterexample (using Min) and apply Amberg's theorem to obtain a maximal subgroup of infinite index. ▯

As a final application we will prove an interesting theorem of Černikov [16], [22].

Theorem 3.45. A locally soluble group which satisfies Min-*ab* is a Černikov group.

We shall first establish

Lemma 3.46. Let G be a locally finite group and let p be a prime. Suppose that X/Y is a section* of G and that $Y < X_1 < X_2 < \cdots$ is an infinite ascending chain of subgroups of X with each X_i/Y a finite p-group. Then there exists a p-subgroup P of X and an ascending chain $P_1 < P_2 < \cdots$ of subgroups of P such that $X_i/Y \simeq P_i/P_i \cap Y$.

Proof. Since X_i/Y is finite and G is locally finite, there is a chain of finite subgroups of X, say $F_1 \leqq F_2 \leqq \cdots$, such that $X_i = YF_i$. For each i we choose a Sylow p-subgroup P_i of F_i in such a way that $P_1 \leqq P_2 \leqq \cdots$. Denote the union of the P_i by P. Now $(Y \cap F_i)\, P_i/Y \cap F_i$ is a Sylow p-subgroup of $F_i/Y \cap F_i$ and the latter is a p-group since it is isomorphic with X_i/Y. Hence $F_i = (Y \cap F_i)\, P_i$ and

$$X_i/Y = YF_i/Y = YP_i/Y \simeq P_i/P_i \cap Y;$$

this of course implies that $P_i < P_{i+1}$. ▯

* A *section* of a group is a factor group of one of its subgroups.

Corollary. Let G be a locally finite group. If G satisfies the minimal condition on abelian p-subgroups, then so does every section of G. If G has finite Sylow p-subgroups, then so has every section of G.

Proof. Let G satisfy the minimal condition on abelian p-subgroups. If P is a p-subgroup of G, then P satisfies Min-*ab* and is a Černikov group by Theorem 3.32. Hence an elementary abelian p-section of P has boundedly finite order; the lemma shows that G can have no infinite elementary abelian p-sections X/Y. Therefore every section inherits the hypothesis on G. The second statement is true for similar reasons. $\square$

Proof of Theorem 3.45. In the proof we are going to use the fact (due to Mal'cev) that a chief factor of a locally soluble group is abelian. For an elementary proof see Theorem 5.27.

Let G be a locally soluble group with Min-*ab* and let $\mathscr{S}$ be a chief series of G. We know that a factor of $\mathscr{S}$ is periodic, abelian and, of course, characteristically simple; hence it is an elementary abelian p-group. By the Corollary to Lemma 3.46, each factor of $\mathscr{S}$ satisfies Min and must therefore be finite. The class

$$\mathfrak{F}^{-\boldsymbol{H}}$$

of groups which have no non-trivial finite homomorphic images is clearly $\boldsymbol{N}$, $\boldsymbol{H}$ and $\boldsymbol{P}$-closed, so it is a radical class. Let R denote the $\mathfrak{F}^{-\boldsymbol{H}}$-radical of G. Since $R \in \mathfrak{F}^{-\boldsymbol{H}}$, we see that R must induce the trivial automorphism in each factor of the series $\mathscr{S}$. By intersecting R with $\mathscr{S}$, term by term, we obtain a central series of R, so R is a Z-group.

Finitely generated subgroups of R are finite and hence nilpotent, which shows that R is locally nilpotent; by Theorem 3.32 the group R is a Černikov group. G/R has no subgroups of type p^∞ for any p, by definition of R, and by the first step of the proof of Theorem 3.32 we see that G/R satisfies Min-*ab*. It follows that each abelian subgroup of G/R is finite. But G/R is locally finite, so it is finite by the Hall-Kulatilaka-Kargapolov Theorem. $\square$

Recently Černikov [35] has shown that *a locally soluble group satisfying the minimal condition on non-normal abelian subgroups is abelian-by-finite.* This evidently implies Theorem 3.45, as does the theorem of Šunkov mentioned on p. 98.

It is an open question whether every locally soluble group satisfying Max-*ab* is polycyclic: in this connection see Gorčakov [4] and Merzljakov [2].

Finally, observe that in proving Theorem 3.45 the Hall-Kulatilaka-Kargapolov Theorem need only be applied to locally soluble groups: consequently the Feit-Thompson Theorem is not required in the proof.

Chapter 4

Finiteness Conditions on Conjugates and Commutators

4.1 Central-by-Finite Groups and Applications of Schur's Theorem

A group is said to be *central-by-finite* if its centre has finite index. The following important result is essentially due to Schur [1]; see also Baer [14], p. 177 and P. Hall [6], Theorem 8.1.

Theorem 4.11. Let N be a subgroup of the centre of a group G and suppose that N has finite index n in G. Then the map $x \to x^n$ is a homomorphism of G into N.

Proof. The proof is a simple application of the transfer. Observe first that $N \lhd G$. If $g \in G$, right multiplication by g induces a permutation $\pi(g)$ of the (finite) set of cosets of N in G. Let the cycle decomposition of $\pi(g)$ be

$$\pi(g) = (X_1 \cdots X_{n_1})\,(Y_1 \cdots Y_{n_2}) \cdots (Z_1 \cdots Z_{n_k})$$

where $n = n_1 + n_2 + \cdots + n_k$: thus for example $X_i g = X_{i+1}$ where $X_{n_1+1} = X_1$. If C is a coset of N, choose any representative element $r(C)$ from C. Then the transfer of G into N is the mapping σ defined by

$$g^\sigma = \prod r(C)\, g r(Cg)^{-1} \tag{1}$$

where the product is taken over all cosets C of N. It is well-known that g^σ is independent of both the order of the factors in the product (1) and the representative function r, and that σ is a homomorphism of G into N (see for example M. Hall [2], § 14.2). Observe that

$$r(C)\, g r(Cg)^{-1} \in N \leq \zeta(G).$$

Now

$$\prod_{i=1}^{n_1} r(X_i)\, g r(X_i g)^{-1} = \prod_{i=1}^{n_1} r(X_i)\, g r(X_{i+1})^{-1} = r(X_1)\, g^{n_1} r(X_1)^{-1},$$

belongs to N, so

$$r(X_1) g^{n_1}\, r(X_1)^{-1} = g^{n_1}.$$

Hence

$$g^\sigma = g^{n_1 + \cdots + n_k} = g^n. \quad \square$$

Possibly the most useful application of this result is

Theorem 4.12. If G is a group whose centre has finite index n, then G' is finite and $(G')^n = 1$.

Once again this is implicit in the work of Schur ([2], p. 26). Other proofs have been given by Baer ([11], p. 357 and [14], pp. 163 and 177), B. H. Neumann ([7], Theorem 5.3), Witt [2], P. Hall ([6], Theorem 8.1) and Kaplansky ([2], Theorem 8.19). Here we present Hall's proof.

Proof of Theorem 4.12. Let $Z = \zeta(G)$. Since the commutator $[x, y]$ depends only on the cosets Zx and Zy, the group G' can be finitely generated (by at most $\binom{n-1}{2}$ elements). $G'/G' \cap Z$ is finite, so Theorem 1.41 shows that $G' \cap Z$ is finitely generated. Also $G' \cap Z$ is abelian; hence it is enough to prove that $(G')^n = 1$. By Theorem 4.11 the mapping $x \to x^n$ is a homomorphism of G into Z; consequently G' is contained in the kernel and $(G')^n = 1$. ▯

Corollary. Let G be a group such that $G/\zeta(G)$ is a locally finite π-group where π is a set of primes. Then G' is a locally finite π-group.

Bounds for the Order of G' when G is Central-by-Finite

Let G be central-by-finite and let $Z = \zeta(G)$ and $n = |G:Z|$. It is of interest to obtain upper bounds for $|G'|$ in terms of n. Since any group of order n can be generated by $[\log_p n]$ elements, where p is the smallest prime dividing n, we can find a subgroup X generated by at most $[\log_p n]$ elements such that $G = XZ$. Now $G' = X'$, so $G' \cap Z \leq X \cap Z$ and $X \cap Z$ needs at most $n([\log_p n] - 1) + 1$ generators, by the Reidemeister-Schreier Theorem. Since $(G')^n = 1$, we obtain the crude bound

$$|G'| \leq n^{n([\log_p n]-1)+2} \tag{2}$$

However in [6] Wiegold has given a method of obtaining sharper upper bounds, which are indeed close to being best possible. For example he shows that *if* $|G: \zeta(G)| = p^m$, *where p is a prime, then*

$$|G'| \leq p^{\frac{1}{2}m(m-1)}. \tag{3}$$

Since Wiegold's proof is very short, we will reproduce it here. If $m \leq 1$, then G is abelian and the assertion is obviously true, so let us assume that $m \geq 2$. Denote by xZ a non-trivial element of the centre of G/Z—here as before $Z = \zeta(G)$. Then $[G, x] \leq Z$; let $H = G/[G, x]$. Now $x[G, x] \in \zeta(H)$, so $Z/[G, x]$ is properly contained in $\zeta(H)$ and $|H: \zeta(H)| = p^l$ where $l < m$. By induction on m we have $|H'| \leq p^{\frac{1}{2}(m-1)(m-2)}$. Next we notice that $[G, x]$ actually *consists* of commutators of the form $[g, x]$, $g \in G$, because $[g, x] \in Z$. Hence

$$|[G, x]| = |G: C_G(x)| \leq p^{m-1}$$

since $C_G(x) \geqq \langle x, Z\rangle > Z$. Finally, $H' = G'/[G, x]$ and it follows that $|G'| \leqq p^{\frac{1}{2}(m-1)(m-2)} \cdot p^{m-1} = p^{\frac{1}{2}m(m-1)}$. For a comparable result about torsion-free nilpotent groups see Bačurin [6].

Wiegold also shows by example that the bound (3) is best possible. On the basis of (3) and Schur's theory of the multiplicator (Schur [2]), Wiegold goes on to prove that *in the general case where* $|G: \zeta(G)| = n$

$$|G'| \leqq n^{\frac{1}{2}\cdot(\log_p n-1)}$$

where p *is the least prime dividing* n. This bound is best possible when $n = p^m$, is a definite improvement on (2) and also improves previous bounds found by Wiegold in [1].

Finiteness Criteria for Commutator Groups. Theorems of Baer

The next result was found by Baer, but with a less exact bound for the exponent ([14], Satz 2 and p. 177).

Theorem 4.13. Let A be a finite group of automorphisms of a group G and let the exponent of A be e. Assume also that $C_G(A^G)$ has finite index n in G. Then $[G, A]$ is finite and its exponent divides ne.

Proof. Let $C = [G, A]$ and $I = C \cap C_G(A^G)$; then $I \leqq \zeta(C)$ since $C \leqq A^G$. The index of I in C divides n, so C is central-by-finite; clearly C is finitely generated. Therefore it will be enough to prove that $C^{ne} = 1$. By Theorem 4.11 the map σ under which $x \to x^n$ is a homomorphism of C into I; moreover, both C and I are A^G-admissible and σ is clearly a A^G-operator homomorphism. Therefore the trivial action of A^G on I implies that A^G has trivial action on $C/\mathrm{Ker}\,\sigma$, and from this it follows that

$$[C, A^G]^n = 1. \tag{4}$$

Next, $C' \leqq [C, A^G] \leqq C$, so $C/[C, A^G]$ is abelian and, of course, it is generated by the elements $[g, \alpha]\,[C, A^G]$, $g \in G$, $\alpha \in A$. Since A has exponent e, we have

$$[g, \alpha]^e \equiv [g, \alpha^e] \equiv 1 \bmod [C, A^G],$$

which shows that

$$C^e \leqq [C, A^G]. \tag{5}$$

From (4) and (5) we conclude that $C^{ne} = 1$. □

Corollary (cf. Baer [14], Satz 2″ and p. 177). Let H, K, M and N be normal subgroups of a group such that $M \leqq H$ and $N \leqq K$. Assume that the indices $|H: M| = h$ and $|K: N| = k$ are finite and also that $[H, N] = 1 = [K, M]$. Then $[H, K]$ is finite and its exponent divides hk.

To prove this we apply Theorem 4.13 with $G = H$ and $A = \mathrm{Aut}_K H$; since $C_K(H) \geqq N$, the group A is finite with exponent dividing k. Also $A^G = A[G, A] = A[H, K]$, so $C_G(A^G) \geqq M$ and $|G: C_G(A^G)|$ is finite and divides h.

When $H = K$ and $M = \zeta(H) = N$, the corollary reduces to Theorem 4.12, but gives a higher exponent for the derived subgroup.

In his paper [14] Baer also considers the following interesting problem: *given two subgroups of a group H and K, when does the finiteness of the set of commutators* $\{[h, k]: h \in H, k \in K\}$ *imply the finiteness of the commutator group* $[H, K]$? Some additional conditions are certainly necessary for an affirmative answer: for example, let $F = H * K$ be the free product of two groups H and K of order 2, so F is an infinite dihedral group; here there is just one non-trivial commutator of the form $[h, k]$, $h \in H$, $k \in K$, but $[H, K]$ is cyclic of infinite order. Baer showed that a sufficient additional condition is the normality of either H of K ([14], Satz 3'); another proof of Baer's theorem together with an application to algebraic groups is in a paper Rosenlicht [1]. Somewhat more generally we will prove

Theorem 4.14. Let H and K be subgroups of a group and suppose that the set of commutators $\{[h, k]: h \in H, k \in K\}$ is finite. Then $[H, K]$ is finite provided that either H or K is ascendant in $\langle H, K\rangle$.

Proof. Let $J = \langle H, K\rangle$ and assume that K *asc* J. Clearly we may suppose that H (but not that K) is finitely generated. Let $h \in H$: the set $\{[h, k]: k \in K\}$ is finite, so there are only finitely many conjugates of h in K and $|K: C_K(h)|$ is finite. Now H is finitely generated and the intersection of the centralizers of the generators of H in K is just $C_K(H)$; therefore $|K: C_K(H)|$ is finite and there is a normal subgroup E of K such that $E \leqq C_K(H)$ and $|K: E|$ is finite. Let $K = \langle k_1, \ldots, k_n, E\rangle$: just as before the finiteness of the set of commutators of the form $[h, k]$ implies that each $|H: C_H(k_i)|$ is finite. Since $[H, E] = 1$, it follows that $|H: C_H(K)|$ is finite and there is a normal subgroup D of H such that $D \leqq C_H(K)$ and $|H: D|$ is finite. Notice that $D \lhd J$ and $E \lhd J$.

Now let

$$\bar{H} = H^K \quad \text{and} \quad \bar{K} = K^H .$$

$\bar{K}/E$ is generated by finitely many subgroups of the form K^x/E, $x \in H$, because K has only a finite number of conjugates in H; also K^x/E is an ascendant finite subgroup of J/E. Therefore $\bar{K}/E$ is locally finite, by Theorem 2.31, and since $\bar{K}/E$ is clearly finitely generated, it is finite. It follows that H has only finitely many conjugates in $\bar{K}$. Now $J = H\bar{K}$, so H/D is a finite subgroup with a finite number of conjugates in J/D and Corollary 3 to Lemma 2.14 shows that $\bar{H}/D$ is finite. $E \lhd J$ implies that $C_J(E) \lhd J$ and therefore $H \leqq C_J(E)$ implies that $\bar{H} \leqq C_J(E)$; in a similar way $\bar{K} \leqq C_J(D)$. We can now apply the Corollary to Theorem 4.13 with $\bar{H}$ for H, $\bar{K}$ for K, D for M and E for N; we conclude that $[\bar{H}, \bar{K}]$ is finite. Hence $[H, K]$ is finite. □

Theorem 4.15 (Baer [14], p. 167). Let A be a group of automorphisms of a group G. Then the finiteness of any two of the following implies the finiteness of the third:

$$\{[g, \alpha]: g \in G, \alpha \in A\}, \quad |G: C_G(A)|, \quad A.^*$$

Proof. If both $|G: C_G(A)|$ and A are finite, it is clear that there are only finitely many $[g, \alpha]$ with $g \in G$ and $\alpha \in A$. Suppose that the set of all $[g, \alpha]$ is finite; then, for all $g \in G$, the index $|A: C_A(g)|$ is finite. If in addition $|G: C_G(A)|$ is finite, let $\{g_1, \ldots, g_n\}$ be a right transversal to $C_G(A)$ in G and observe that the intersection of the $C_A(g_i)$ is trivial. Hence A is finite. Finally let the set of all $[g, \alpha]$ be finite and let A be finite. Since $G \lhd GA$, Theorem 4.14 implies that $[G, A]$ is finite. Hence $A^G = [G, A]A$, and therefore $|G: C_G(A^G)|$, is finite. $C_G(A^G) \leqq C_G(A)$, so $|G: C_G(A)|$ is finite. □

Turner-Smith has examined in greater detail the consequences of the finiteness of the set $\{[g, \alpha]: g \in G, \alpha \in A\}$; for example, it is easy to show that A and $G/C_G(A^G)$ are abelian-by-finite and have finite exponents ([1], Theorem 3.1). Also, if $|G: C_G(A)|$ is finite, $[G, A]$ is central-by-finite and A is abelian-by-finite ([1], Theorem 3.3).

Groups with Finite Coverings of Abelian Subgroups

A group is said to be *covered* by a collection of subsets if each element of the group belongs to at least one subset in the collection: the collection of subsets is called a *covering* of the group.

In an unpublished result Baer has characterized the central-by-finite groups in terms of coverings: (see B. H. Neumann [11], pp. 227—228).

Theorem 4.16. A group is central-by-finite if and only if it has a finite covering consisting of abelian subgroups.

To prove this we will employ a lemma due to B. H. Neumann which is useful in studying coverings by cosets.

Lemma 4.17 (B. H. Neumann [9], 4.4). Let $G = \bigcup_{i=1}^{n} S_i g_i$ where $S_1, \ldots, S_n$ are (not necessarily distinct) subgroups of G. Then we can omit from the union any $S_i g_i$ for which $|G: S_i|$ is infinite.

* Notice that $C_G(A)$ is the set of fixed points of the automorphism group A.

Proof. We show first that *at least one S_i has finite index in G*. Suppose that there are r distinct subgroups among $S_1, \dots, S_n$; if $r = 1$, then G is the union of finitely many right cosets of S_1 and $|G : S_1|$ is finite. Assume therefore that $r > 1$ and that the S_i are labelled so that $S_{m+1} = \cdots = S_n$ and S_n is distinct from each of $S_1, \dots, S_m$. If

$$G = \bigcup_{i=m+1}^{n} S_n g_i,$$

then $|G : S_n|$ is finite, so we may assume that there is an element $x \notin \bigcup_{i=m+1}^{n} S_n g_i$. Then

$$(S_n x) \cap \left(\bigcup_{i=m+1}^{n} S_n g_i \right)$$

is empty and consequently

$$S_n x \subseteq \bigcup_{i=1}^{m} S_i g_i.$$

It follows that for $m + 1 \leqq j \leqq n$

$$S_n g_j \subseteq \bigcup_{i=1}^{m} S_i g_i x^{-1} g_j.$$

Therefore G is the union of finitely many right cosets of $S_1, \dots, S_m$. Only $r - 1$ of the latter are distinct, so by induction on r at least one of them has finite index.

We can assume that there are subgroups of infinite index among the S_i and that the S_i are so ordered that $S_1, \dots, S_l$ are all of infinite index and $S_{l+1}, \dots, S_n$ are all of finite index. By the first part of the proof $l < n$. Let $I = S_{l+1} \cap \cdots \cap S_n$; then I has finite index in G and each right coset of S_j, where $l + 1 \leqq j \leqq n$, is the union of finitely many right cosets of I. Hence G is the union of finitely many right cosets of $S_1, \dots, S_l$ and I. If G is the union of the right cosets of I only, then we can restore the original right cosets of $S_{l+1}, \dots, S_m$, with the result that the cosets of $S_1, \dots, S_l$ have been eliminated from the union. Otherwise the argument of the last paragraph, with l instead of m, shows that one of $S_1, \dots, S_l$ has finite index in G, which is a contradiction. ▯

Proof of Theorem 4.16. Suppose that G is central-by-finite and let $Z = \zeta(G)$: choose a transversal to Z in G, say $\{g_1, \dots, g_n\}$. Then for any $g \in G$, we can write $g = g_i z$ where $z \in Z$. Hence $g \in \langle g_i, Z \rangle$, which is abelian, and G is covered by the $\langle g_i, Z \rangle$ with $i = 1, \dots, n$.

Conversely assume that G is covered by finitely many abelian subgroups; then Lemma 4.17 shows that G is covered by finitely many abelian subgroups of finite index; let I be their intersection. Then

$|G: I|$ is finite and if $g \in G$, both g and I are contained in an abelian subgroup, so $[g, I] = 1$ and $I \leq \zeta(G)$. □

In [11] B. H. Neumann has obtained bounds for the index $|G: \zeta(G)|$ in terms of the minimal number of subgroups in a finite covering of G by abelian subgroups; his methods are applicable to the more general case of a group covered by finitely many cosets. We mention also that B. H. Neumann has shown that *a group G can be covered by permutable, boundedly finite subsets if and only if* $G \in \mathfrak{F}\mathfrak{A}\mathfrak{F}$ ([9], Theorem 7.1).

Finally we shall give another unpublished covering theorem due to Baer.

Theorem 4.18. The hypercentre of a group G has finite index if and only if G has a finite covering consisting of hypercentral subgroups.

Proof. Suppose that G has a finite covering consisting of hypercentral subgroups. By Lemma 4.17 we can assume that each member of this covering has finite index in G. First we form the intersection of all members of the covering and then we replace this by its core in G, say N. Then N is a normal hypercentral subgroup with finite index in G which is contained in every member of the covering. It is enough to prove that N lies in the hypercentre of G. Suppose that this is not the case; then, by passing to $G/\bar{\zeta}(G)$, we can assume that $\bar{\zeta}(G) = 1$ and $N \neq 1$.

Let $Z = \zeta(N)$; then $Z \neq 1$ because N is hypercentral. Choose a nontrivial element a from Z. Now $N \leq C_G(a)$, so $|G: C_G(a)|$ is finite and consequently a^G is a finitely generated abelian group. Hence we can find a nontrivial normal subgroup A of G which is contained in a^G and is either a free abelian group of finite rank or a finite elementary abelian p-group for some prime p. Clearly $N \leq C_G(A)$ and $X = G/C_G(A)$ is finite. The action of elements of G on A by conjugation enables us to represent X by a group of matrices over either the ring of integers or the field $GF(p)$. Let $g \in G$; now $\langle g, A\rangle$ is contained in some member of the covering, so it is hypercentral. Also $\langle g, A\rangle$ is finitely generated, so it is even nilpotent. It follows that it is possible to represent $gC_G(A)$ by a unitriangular matrix by choosing a suitable basis for A. Such a matrix must in our case have order either ∞ or a power of p. Hence X is is either torsion-free or a p-group. In the first case X has to be trivial and $1 \neq A \leq \zeta(G)$, contrary to assumption. Hence A and X are finite p-groups. The semidirect product S of A by X, formed in the obvious manner, is also a finite p-group, so it is nilpotent. From Lemma 2.16 we obtain $A \cap \zeta(S) \neq 1$, which is equivalent to $A \cap \zeta(G) \neq 1$, our final contradiction.

The proof of the converse implication is straightforward and is left to the reader. □

Groups which Have Finite Automorphism Groups

The groups which are central-by-finite are just the groups whose inner automorphism groups are finite. This raises the problem of describing the groups whose automorphism group is finite. The simplest example of an infinite group with finite automorphism group is, of course, an infinite cyclic group, which has automorphism group of order 2. In fact it is known that there exist uncountable torsion-free abelian groups with automorphism group of order 2—here we refer to papers of de Groot [1], Fuchs [2], Hulanicki [1] and Saşiada [1]—and it seems likely that the task of finding all groups with finite automorphism group is a very difficult one. However, when such groups are subjected to suitable additional conditions, interesting results can be obtained.

Theorem 4.19 (Alperin [1]). A finitely generated group has finite automorphism group if and only if it is either finite or a finite extension of an infinite cyclic group lying in its centre.

Theorem 4.19.1. An infinite group which has finite automorphism group possesses infinitely many monendomorphisms* which map each element to one of its powers.

Corollary 1. A group which has only a finite number of monendomorphisms finite.

In particular *a group with finitely many endomorphisms is finite* (Baer [23], p. 530).

Corollary 2 (Baer [23], p. 529). A periodic group with finite automorphism group is finite.

For, in a periodic group a monendomorphism under which each element is mapped to a power is an automorphism.

The proofs of these theorems depend on the following remark. Let G be a central-by-finite group, let $Z = \zeta(G)$ and let $|G:Z| = n$. Denote by α any endomorphism of Z and observe that $x^{n\alpha} \equiv (x^n)^\alpha \in Z$ for all $x \in G$. Let $\theta(\alpha)$ be the mapping of G into itself under which

$$x \to x^{1+n\alpha} = x(x^n)^\alpha.$$

Since $x \to x^n$ is homomorphic, by Theorem 4.11, we find that

$$(xy)^{\theta(\alpha)} = xy((xy)^n)^\alpha = xyx^{n\alpha}y^{n\alpha} = x^{1+n\alpha}y^{1+n\alpha} = x^{\theta(\alpha)}y^{\theta(\alpha)},$$

using the fact that $x^{n\alpha}$ belongs to the centre of G. Hence $\theta(\alpha)$ is an endomorphism of G. Clearly $Z^{\theta(\alpha)} \leq Z$. If $x^{\theta(\alpha)} = 1$, then $x = (x^{n\alpha})^{-1} \in Z$;

---

* A *monendomorphism* is an endomorphism which is also a monomorphism.

thus $\theta(\alpha)$ *is a monomorphism if and only if its restriction to Z is a monomorphism.* Also for any $x \in G$ we may write

$$x = x^{\theta(\alpha)}(x^{n\alpha})^{-1} \equiv x^{\theta(\alpha)} \bmod Z;$$

thus $\theta(\alpha)$ *is an epimorphism if and only if its restriction to Z is an epimorphism.*

Proof of Theorem 4.19 (Scott [5], Theorem 15.1.19). Let G be a finitely generated infinite group with finite automorphism group and let $Z = \zeta(G)$. Then $n = |G:Z|$ is finite, and Z is finitely generated, by Theorem 1.41. Thus Z is the direct product of finitely many cyclic groups at least one of which must be infinite. Suppose that

$$Z = \langle a \rangle \times \langle b \rangle \times Y$$

where $\langle a \rangle$ and $\langle b \rangle$ are infinite cyclic groups. For each integer i let α_i be the endomorphism of Z such that

$$a^{\alpha_i} = b^i, \quad b^{\alpha_i} = 1 \quad \text{and.} \quad y^{\alpha_i} = 1 \quad (y \in Y).$$

The map $\theta(\alpha_i)$ in which $x \to x^{1+n\alpha_i}$ is an endomorphism of G. Now the restriction of $\theta(\alpha_i)$ to Z maps a to ab^{ni}, b to b and y to y, $(y \in Y)$, and this is an automorphism of Z since it has an inverse, namely the mapping under which $a \to ab^{-ni}$, $b \to b$ and $y \to y$, $(y \in Y)$. Hence $\theta(\alpha_i)$ is an automorphism of G by the remarks preceding this proof. Since b has infinite order, $\theta(\alpha_i) \neq \theta(\alpha_j)$ if $i \neq j$, and Aut G is infinite. It follows that exactly one of the cyclic direct factors of Z is infinite and hence that G is a finite extension of this cyclic subgroup.

Conversely, let G have this structure and let A = Aut G. There is a positive integer m such that G^m is an infinite cyclic subgroup of $\zeta(G)$. Now G/G^m is finite and G^m is characteristic in G. Hence the automorphism groups induced by A in G^m and G/G^m are finite. Also Hom $(G/G^m, G^m) = 0$, which is easily seen to imply that only the identity automorphism of G can act trivially on both G^m and G/G^m. Hence A is finite. □

Proof of Theorem 4.19.1. Let G be an infinite group with finite automorphism group and let $Z = \zeta(G)$ and $n = |G:Z|$. We begin by showing that *G is not periodic* (which is the content of Corollary 2). Suppose that G is periodic. If p is a prime, let Z_p denote the p-component of Z. The mappings $a \to a^{p^i}$, $(a \in Z_p)$, and $a \to 1$, $(a \in Z_q,\ q \neq p)$, determine an endomorphism $\alpha_{p,i}$ of Z and, if $i > 0$, then $\theta(\alpha_{p,i})$ is an automorphism of G by the remarks preceding these proofs. Since Aut G is finite, $\theta(\alpha_{p,i}) = \theta(\alpha_{p,j})$ for some $i \neq j$, which implies that Z_p has finite exponent. Thus Z is a direct product of cyclic groups (see Kaplansky [1], p. 17, Theorem 6). Now there is a finitely generated subgroup X such that $G = XZ$, and X is even finite because G is locally finite. Hence there is a decomposition $Z = D \times E$ where D is finite and contains $X \cap Z$. Obviously $Y = XD$ is finite. Now X, D, E, and hence Y, are normal in G and

$$G = XZ = XDE = YE;$$

also

$$Y \cap E = (XD) \cap Z \cap E = ((X \cap Z)\, D) \cap E = D \cap E = 1.$$

It follows that $G = Y \times E$. From this we deduce that $\operatorname{Aut} E$ is isomorphic with a subgroup of $\operatorname{Aut} G$ and is therefore finite. But E is a direct product of cyclic groups, so it is clear that E, and hence G, is finite. This is a contradiction.

Suppose next that Z contains elements of only finitely many different prime orders. By Dirichlet's theorem on the infinity of primes in an arithmetic progression (Hasse [2], p. 167) there exist infinitely many primes of the form $p = 1 + nr$ such that Z contains no elements of order p. Now $x \to x^r$ is an endomorphism of Z and the map $\theta_r: x \to x^{1+nr} = x^p$ is an endomorphism of G which, by choice of p, is monomorphic on Z and hence on G. If $\theta_r = \theta_s$ where $r \neq s$, then Z has finite exponent and G is periodic. Hence the θ_r are all distinct and the theorem is proved in this case.

From now on we assume that Z has elements of infinitely many distinct prime orders. Denote by π the set of prime divisors of $2n$ and let P be the π-component of Z; then Z/P is not torsion-free. A well-known theorem in the theory of abelian groups (Kaplansky [1], p. 21, Theorem 9) allows us to conclude that Z/P has a non-trivial direct factor, say D/P, which is either a p^∞-group or a cyclic group of order p^l, where p is a prime. Suppose that in fact $Z = DE$ where $D \cap E = P$. Now P is a direct factor of D because $p \notin \pi$; let $D = P \times F$. Then

$$Z = DE = PFE = FE$$

and

$$F \cap E = F \cap D \cap E = F \cap P = 1.$$

Therefore

$$Z = F \times E \tag{6}$$

and F is either of type p^∞ or is cyclic of order p^l.

Suppose that F is of type p^∞ and let λ be a p-adic integer such that $\lambda \equiv 0 \bmod p$. The map $f \to f^\lambda$—where λ is reduced modulo the order of f—is an endomorphism of F which extends to an endomorphism α of Z such that $e^\alpha = 1$ if $e \in E$, by (6). By our choice of λ we have

$$1 + n\lambda \not\equiv 0 \bmod p$$

thus the map $\theta(\alpha): x \to x^{1+n\alpha}$ is an automorphism of G. Clearly different choices of λ lead to different $\theta(\alpha)$ and there are infinitely many λ's available, so we have a contradiction.

Finally, let F have order p^l where $l > 0$. By construction p does not divide $2n$. It follows that there is an integer t such that $1 + nt \not\equiv 0$ or $1 \bmod p$ and $0 < t < p$. Let α_1 be the endomorphism of Z under which $f \to f^t$ and $e \to 1$, $(f \in F, e \in F)$. Then $\theta(\alpha_1): x \to x^{1+n\alpha_1}$ is a non-identity automorphism of G, by choice of t; also $C_Z(\theta(\alpha_1)) = E$. Now E/P is not torsion-free, so the same argument can be applied to E and we obtain $E = F_1 \times E_1$ and $Z = F \times F_1 \times E_1$ where F_1 is either quasicyclic or cyclic of non-trivial prime power order; naturally we can assume that it is the latter. Then there is an endomorphism α_2 of Z such that $\theta(\alpha_2)$ is an automorphism of G and $C_Z(\theta(\alpha_2)) = F \times E_1$, so that $\theta(\alpha_1) \neq \theta(\alpha_2)$. Clearly we can keep repeating this procedure and generate an infinite sequence of distinct automorphisms $\theta(\alpha_1), \theta(\alpha_2), \ldots$ of G. □

Theorem 4.19.2. Every homomorphic image of a group G has finite automorphism group if and only if G is either finite or a finite extension of an infinite cyclic group lying in its centre.

Proof. Let G be an infinite group such that every homomorphic image of G has finite automorphism group. If $Z = \zeta(G)$, then $|G:Z|$ is certainly finite. If Z were periodic, G would also be periodic and hence finite, by Corollary 2 to Theorem 4.19.1. Consequently Z contains an element of infinite order. By Zorn's Lemma there exists a subset S of Z which is maximal with respect to being linearly independent and containing only elements of infinite order. Evidently S generates a free abelian subgroup F of Z and G/F is periodic. Since $\mathrm{Aut}(G/F)$ is finite, G/F is finite. If F is not infinite cyclic, we can find a subgroup F_1 such that F/F_1 is a free abelian group of rank 2. Now G/F_1 is a finitely generated group with finite automorphism group, so Theorem 4.19 is applicable. Thus we are led to the contradiction that F/F_1 is infinite cyclic. The converse follows from Theorem 4.19. □

In conclusion we mention that de Vries and de Miranda [1] have determined all groups of order ≤ 8 which can occur as the automorphism group of some group. Hallett and Hirsch [1], [2], and also Hirsch and Zassenhaus [1], have determined all finite groups which can occur as the automorphism group of a torsion-free group.

Groups whose homomorphic images have countable automorphism groups are the subject of an extensive paper of Baer [52]; see also Schlette [1].

4.2 Three Problems of Philip Hall on Verbal and Marginal Subgroups

Let θ be a word in n variables and let G be a group: we recall that $\theta(G)$ and $\theta^*(G)$ are the verbal and marginal subgroups of G determined by θ. The following problems are generally attributed to P. Hall (cf. P. Hall [6], § 8 and Turner-Smith [1], p. 321):

(I) *If π is a set of primes and $|G: \theta^*(G)|$ is a finite π-group, is $\theta(G)$ also a finite π-group?*

(II) *If $\theta(G)$ is finite and G satisfies Max, is $|G: \theta^*(G)|$ finite?*

(III) *If θ is finite-valued on G, that is, if the set $\{\theta(g_1, \ldots, g_n): g_i \in G\}$ is finite, does it follow that $\theta(G)$ is finite?*

If $\theta(x_1, x_2) = [x_1, x_2]$, then $\theta(G) = G'$ and $\theta^*(G) = \zeta(G)$; thus (I) and (III) have positive answers in this case, by Theorems 4.12 and 4.14 respectively. In fact (II) also has a positive answer for this θ (see Theorem 4.24).

When θ is an arbitrary word, none of these problems has been settled. However there has been a good deal of progress in recent years, thanks to the work of P. Hall, Merzljakov, Stroud and Turner-Smith. It is our intention to give an account of some of the results obtained by these authors.

Halls's First Problem. Schur Pairs

If θ is a word and $\mathfrak{X}$ is a class of groups, we will call

$$(\theta, \mathfrak{X})$$

a *Schur pair* if $G/\theta^*(G) \in \mathfrak{X}$ always implies that $\theta(G) \in \mathfrak{X}$. Thus Problem (I) asks whether $(\theta, \mathfrak{F}_\pi)$ is always a Schur pair.

If θ and ϕ are arbitrary words in m and n variables respectively, we define a word

$$\psi = [\theta, \phi]$$

in $m + n$ variables by the rule

$$\psi(x_1, \ldots, x_m, y_1, \ldots, y_n) = [\theta(x_1, \ldots, x_m), \phi(y_1, \ldots, y_n)].$$

Since $[\theta(G), \phi(G)]$ is generated by all conjugates of values of ψ in G (by Lemma 2.12), and hence by all values of ψ in G, it follows that

$$\psi(G) = [\theta(G), \phi(G)].$$

A word θ is called a *commutator word* if the inclusion

$$\theta(G) \leq G'$$

is always valid. There are, of course, many types of commutator words: among the most accessible of these are the *outer commutator words* defined as follows. The unique outer commutator word of weight 1 is γ_1 where $\gamma_1(x_1) = x_1$; and θ is an outer commutator word of weight $w > 1$ if $\theta = [\theta_1, \theta_2]$ where θ_i is an outer commutator word of weight w_i and $w = w_1 + w_2$.

Foremost among the outer commutator words are the *lower central words* γ_n defined by

$$\gamma_n = [\gamma_{n-1}, \gamma_1], \quad (n > 1);$$

clearly $\gamma_n(x_1, \ldots, x_n) = [x_1, \ldots, x_n]$.

In [1] Stroud, generalizing a result of P. Hall ([6], Theorem 8.7), showed that if $(\theta, \mathfrak{X})$ and $(\phi, \mathfrak{X})$ are Schur pairs, then so is $([\theta, \phi], \mathfrak{X})$, provided that $(\gamma_2, \mathfrak{X})$ is a Schur pair and $\mathfrak{X} = \boldsymbol{S}_n\mathfrak{X} = \boldsymbol{H}\mathfrak{X} = \boldsymbol{P}\mathfrak{X} \leqq \mathfrak{G}$. We will now prove a result that is, in fact, a generalization of Stroud's theorem.

Theorem 4.21. Let $\mathfrak{X} = \boldsymbol{S}_n\mathfrak{X} = \boldsymbol{H}\mathfrak{X} = \boldsymbol{P}\mathfrak{X}$ and assume that the tensor product of two abelian $\mathfrak{X}$-groups is an $\mathfrak{X}$-group. Assume also that $(\gamma_2, \mathfrak{X})$ is a Schur pair. If θ and ϕ are words for which $(\theta, \mathfrak{X})$ and $(\phi, \mathfrak{X})$ are Schur pairs, then $([\theta, \phi], \mathfrak{X})$ is a Schur pair.

Corollary 1. If $\mathfrak{X}$ is a class of groups satisfying the hypotheses of the theorem and θ is any outer commutator word, then $(\theta, \mathfrak{X})$ is a Schur pair.

This follows from the theorem by induction on the weight of θ. Setting $\theta = \gamma_{i+1}$ in this corollary we obtain

Corollary 2. If G is a group such that $G/\zeta_i(G) \in \mathfrak{X}$, then $\gamma_{i+1}(G) \in \mathfrak{X}$, where i is finite and $\mathfrak{X}$ is a class of groups satisfying the hypotheses of Theorem 4.21.

By Theorem 4.12 we may take $\mathfrak{X}$ to be $\mathfrak{F}_\pi$; thus Hall's first problem has a positive solution for outer commutator words. When $\mathfrak{X} = \mathfrak{F}_\pi$, Corollary 2 is due to Baer ([11], p. 369, Theorem 4, and [14], p. 173).

Notice that in Corollary 2 the integer i cannot be replaced by an infinite ordinal. For if G is a locally dihedral 2-group, then $|G : \zeta_\omega(G)| = 2$, yet $\gamma_{\omega+1}(G) = G'$ is of type 2^∞.

The proof of Theorem 4.21, which is essentially the same as that of Stroud's original result, depends on the following partial generalization of the Corollary to Theorem 4.13.

Lemma 4.22. Let $\mathfrak{X}$ be a class of groups satisfying the conditions of Theorem 4.21. Let H, K, M and N be normal subgroups of a group such that $M \leqq H$ and $N \leqq K$. Assume that H/M and K/N belong to $\mathfrak{X}$ and also that $[H, N] = 1 = [K, M]$. Then $[H, K]$ belongs to $\mathfrak{X}$.

Proof. Let $I = H \cap K$ and $\bar{K} = K/C_K(I)$. We form the semi-direct product P of I by $\bar{K}$ and note that $[M \cap I, \bar{K}] = 1 = [M \cap I, I]$, so that $M \cap I$ lies in the centre of P. Now $I/M \cap I \simeq IM/M \lhd H/M$ and, by the $\boldsymbol{S}_n$-closure of $\mathfrak{X}$, we have $I/M \cap I \in \mathfrak{X}$. Also $N \leqq C_K(I)$ and hence $\bar{K} \in \boldsymbol{H}\mathfrak{X} = \mathfrak{X}$. Therefore P/I, being isomorphic with $\bar{K}$, belongs to $\mathfrak{X}$ and it follows that $P/\zeta(P) \in \boldsymbol{HP}\mathfrak{X} = \mathfrak{X}$. By hypothesis $(\gamma_2, \mathfrak{X})$ is a Schur pair, so $P' \in \mathfrak{X}$ and $[I, K] = [I, \bar{K}] \in \boldsymbol{S}_n\mathfrak{X} = \mathfrak{X}$. In a similar way we show that $[I, H] \in \mathfrak{X}$, and consequently

$$[I, H][I, K] \in \mathbf{N}_0\mathfrak{X} \leq \boldsymbol{PH}\mathfrak{X} = \mathfrak{X}.$$

Since $[I, H][I, K]$ is a normal subgroup and $\mathfrak{X}$ is $\boldsymbol{P}$-closed, we can pass to the factor group modulo $[I, H][I, K]$ and assume that

$$[I, H] = 1 = [I, K].$$

Since $[H, K] \leq I$, this implies that $[H, K, H] = 1 = [H, K, K]$; hence $[H', K] = 1 = [H, K']$, by the Three Subgroup Lemma. From these four equations and the hypotheses $[H, N] = 1 = [K, M]$ we see at once that the map

$$(hH'M) \otimes (kK'N) \to [h, k], \quad (h \in H, k \in K),$$

determines a well-defined homomorphism of $(H/H'M) \otimes (K/K'N)$ onto $[H, K]$. Since H/M and K/N belong to $\mathfrak{X}$, the hypotheses on $\mathfrak{X}$ imply that $[H, K]$ belongs to $\mathfrak{X}$. □

Proof of Theorem 4.21. $(\theta, \mathfrak{X})$ and $(\phi, \mathfrak{X})$ are Schur pairs: we have to show that $(\psi, \mathfrak{X})$ is a Schur pair where $\psi = [\theta, \phi]$. For the present let G be any group and let $C_\theta = C_G(\theta(G))$ and $C_\phi = C_G(\phi(G))$. Two normal subgroups $N_{\theta,\phi}$ and $N_{\phi,\theta}$ are determined by

$$\theta^*(G/C_\phi) = N_{\theta,\phi}/C_\phi \quad \text{and} \quad \phi^*(G/C_\theta) = N_{\phi,\theta}/C_\theta.$$

We will prove that

$$\psi^*(G) = N_{\theta,\phi} \cap N_{\phi,\theta}, \tag{7}$$

a result which is due to P. Hall ([6], Lemma 8.5). By definition an element a belongs to $\psi^*(G)$ if and only if

$$[\theta(x_1, \ldots, x_i a, \ldots, x_m), \phi(y_1, \ldots, y_n)] = [\theta(x_1, \ldots, x_i, \ldots, x_m), \phi(y_1, \ldots, y_n)]$$

and

$$[\theta(x_1, \ldots, x_m), \phi(y_1, \ldots, y_j a, \ldots, y_n)] = [\theta(x_1, \ldots, x_m), \phi(y_1, \ldots, y_j, \ldots, y_n)]$$

for all x_r and y_s in G and $1 \leq i \leq m$ and $1 \leq j \leq n$. Since $[u, v] = [u, w]$ is equivalent to $[u, vw^{-1}] = 1$, the element a belongs to $\psi^*(G)$ if and only if for all x_r, y_s, i and j,

$$\theta(x_1, \ldots, x_i a, \ldots, x_m)\, \theta(x_1, \ldots, x_i, \ldots, x_m)^{-1} \in C_\phi$$

and

$$\phi(y_1, \ldots, y_j a, \ldots, y_n)\, \phi(y_1, \ldots, y_j, \ldots, y_n)^{-1} \in C_\theta.$$

Hence $a \in \psi^*(G)$ if and only if $a \in N_{\theta,\phi} \cap N_{\phi,\theta}$.

Now suppose that $G/\psi^*(G) \in \mathfrak{X}$. Then $G/N_{\theta,\phi} \in \boldsymbol{H}\mathfrak{X} = \mathfrak{X}$, by (7). Since

$$(G/C_\phi)/\theta^*(G/C_\phi) \simeq G/N_{\theta,\phi}$$

and $(\theta, \mathfrak{X})$ is a Schur pair, $\theta(G)/\theta(G) \cap C_\phi \simeq \theta(G/C_\phi) \in \mathfrak{X}$: by symmetry $\phi(G)/\phi(G) \cap C_\theta \in \mathfrak{X}$. We now apply Lemma 4.22 with $H = \theta(G)$, $K = \phi(G)$ $M = \theta(G) \cap C_\phi$ and $N = \phi(G) \cap C_\theta$ and conclude that

$$\psi(G) = [\theta(G), \phi(G)] \in \mathfrak{X}. \quad \square$$

In his paper [1] Stroud points out that Lemma 4.22, and hence Theorem 4.21, is valid when $\mathfrak{X}$ is the class of finitely generated nilpotent groups, even although this class is not $\boldsymbol{P}$-closed. For, with the notation of the lemma, $[H, K]$ is polycyclic in this case (see below) and

$$[H, K]/[H, K] \cap M \cap N$$

is nilpotent; but $[H, K] \cap M \cap N$ lies in the centre of HK, so $[H, K]$ is nilpotent.

Stroud [1] also proves that *if θ is a word determining a variety all of whose members are locally residually finite, then $(\theta, \mathfrak{F}_\pi)$ is a Schur pair; if the variety determined by θ consists of locally finite groups or of nilpotent groups, then $(\theta, \boldsymbol{P}\mathfrak{C})$ is a Schur pair.* P. Hall has proved that *if G is polycyclic-by-finite and $G/\theta^*(G)$ is finite, then $\theta(G)$ is finite for an arbitrary word θ* ([6], Theorem 8.9).

Some Classes $\mathfrak{X}$ for which $(\gamma_2, \mathfrak{X})$ is a Schur Pair

Classes of groups $\mathfrak{X}$ with the property that $(\gamma_2, \mathfrak{X})$ is a Schur pair are of particular interest, especially if they satisfy the other hypotheses of Theorem 4.21. Some fairly obvious candidates for $\mathfrak{X}$ are

$$\mathfrak{F}_\pi, \quad \boldsymbol{L}\mathfrak{F}_\pi, \quad \boldsymbol{P}\mathfrak{C}, \quad (\boldsymbol{P}\mathfrak{C})\mathfrak{F}.$$

That the first two of these are possibilities follows from Theorem 4.12 and that the other two are eligible is easy to prove directly. For example, let $G/\zeta(G)$ be polycyclic-by-finite and write Z for $\zeta(G)$. We can find a finitely generated subgroup X such that $G = XZ$. Now $X/X \cap Z$ is finitely presented and hence $X \cap Z$ is finitely generated, by Lemma 1.43; therefore X is polycyclic-by-finite. Finally, $G' = X'$ because $[G, Z] = 1$. Hence G' is polycyclic-by-finite.

Less trivial is the following result of Polovickiĭ [4]—see also Schlette [1] (Proposition 3.2).

Theorem 4.23. If $G/\zeta(G)$ is a Černikov group, then G' is a Černikov group.

Proof. Let $Z = \zeta(G)$ and denote by R/Z the finite residual of G/Z. Then G/R is finite and R/Z is a radicable abelian group satisfying Min. The mapping $(Zx, Zy) \rightarrow [x, y]$ induces a homomorphism from $(R/Z) \otimes (R/Z)$ onto R'. But the tensor product of two periodic radicable abelian groups is always trivial, so $R' = 1$ and R is abelian.

Since G/R is finite, we can write $G = RX$ for some finitely generated subgroup X. Clearly $X/X \cap Z$ is finite and $X \cap Z \leq \zeta(X)$, so $X/\zeta(X)$ is finite. Theorem 4.12 shows that X' is finite. Since R is abelian, $G' = [R, X]\, X'$; evidently $[R, X] \lhd G$, so it is enough to show that $[R, X]$ is a Černikov group. Since $X/X \cap Z$ is finite, $[R, X]$ may be expressed as the product of finitely many subgroups of the form $[R, x]$ where $x \in X$. Bearing in mind that $[R, X]$ is abelian and that if $x \in X$, the map $Za \rightarrow [a, x]$ is a homomorphism of R/Z onto $[R, x]$, we conclude $[R, x]$, and hence $[R, X]$, is a Černikov group. ▯

Corollary. Let θ be an outer commutator word. If G is a group for which $G/\theta^*(G)$ is a Černikov group, then $\theta(G)$ is a Černikov group.

This follows from Theorem 4.23 and Corollary 1 to Theorem 4.21.

Hall's Second Problem

Problem (II), like Problem (I), has a positive solution for outer commutator words.

Theorem 4.24 (Turner-Smith [1], pp. 321—322). Let θ be an outer commutator word and let G be group which satisfies Max. If $\theta(G)$ is finite, then $|G: \theta^*(G)|$ is finite.

Proof. Let w be the weight of θ; if $w = 1$, then $\theta(G) = G$ and $\theta^*(G)=1$. Hence we can assume that $w > 1$ and use induction on w; then $\theta = [\theta_1, \theta_2]$ where θ_i has smaller weight than θ. Let $a \in \theta_2(G)$; since $\theta(G)$ is finite, the set of conjugates $\{a^b : b \in \theta_1(G)\}$ is finite, so $C_{\theta_1(G)}(a)$ has finite index in $\theta_1(G)$. Since G satisfies Max, $\theta_2(G)$ is finitely generated and consequently $\theta_1(G)/\theta_1(G) \cap C_{\theta_2}$ is finite; here C_{θ_i} denotes $C_G(\theta_i(G))$. It follows that $\theta_1(G/C_{\theta_2})$ is finite; thus we can apply the induction hypothesis to θ_1 and the group G/C_{θ_2} and conclude that $G/N_{\theta_1,\theta_2}$ is finite, where

$$\theta_1^*(G/C_{\theta_2}) = N_{\theta_1,\theta_2}/C_{\theta_2}.$$

By symmetry $G/N_{\theta_2,\theta_1}$ is finite, where N_{θ_2,θ_1} is defined analogously. But $\theta^*(G) = N_{\theta_1,\theta_2} \cap N_{\theta_2,\theta_1}$, by equation (7) on p. 114. Therefore $\theta^*(G)$ has finite index in G as required. ▯

See Turner-Smith [1] (Theorem 4.1) for a generalization of Theorem 4.24.

It should be observed that if we omit the requirement that G satisfy Max from its statement, Problem (II) has a negative solution, even when $\theta = \gamma_2$. For let G be the central product (i.e. the direct product with amalgamated centres) of an infinite set of copies of the quaternion group of order 8: here $G' = \zeta(G)$ has order 2, but $|G: \zeta(G)|$ is infinite. However P. Hall has proved the following result.

Theorem 4.25 (P. Hall [5]). If G is a group such that $\gamma_{i+1}(G)$ is finite for some integer i, then $|G: \zeta_{2i}(G)|$ is finite.

From this theorem and from Corollary 2 to Theorem 4.21 we derive the following information. *A group G is finite-by-nilpotent if and only if $|G: \zeta_j(G)|$ is finite for some integer $j \geqq 0$.* Also, if $G \in \mathfrak{F}\mathfrak{N}$, we may define two invariants of G, namely $i = i(G)$, the least non-negative integer such that $\gamma_{i+1}(G)$ is finite, and $j = j(G)$, the least non-negative integer such that $|G: \zeta_j(G)|$ is finite, and they satisfy the inequalities

$$i(G) \leqq j(G) \leqq 2i(G).$$

What is more, P. Hall has shown by construction that to every pair of integers i and j such that $0 < i \leqq j \leqq 2i$ there corresponds a nilpotent group G such that $i = i(G)$ and $j = j(G)$. (P. Hall [5]).

To simplify the notation during the proof of Hall's theorem we will write

$$\gamma HK^i = [H, \underbrace{K, \dots, K}_{i}].$$

The proof of the theorem requires the following lemma.

Lemma 4.26 (P. Hall [5]). Let G be any group and let $H = C_G(\gamma_{h+1}(G))$ and $K = C_G(\gamma_{k+1}(G))$ where h and k are integers with $h + k > 0$. Then

$$[H, K] \leqq \zeta_{h+k-1}(G)$$

and also, if $a + b + c \geqq 2h - 1$,

$$[\gamma HG^a, \gamma HG^b] \leqq \zeta_c(G).$$

Proof. If r is any non-negative integer and M and N are arbitrary normal subgroups of G, then

$$\gamma([M, N])\, G^r \leqq \prod_{i+j=r} [\gamma MG^i, \gamma NG^j]; \tag{8}$$

this is easily proved by induction on r with the aid of the Three Subgroup Lemma. Set $r = h + k - 1$ and assume that $i + j = r$. If $i < k$, then $j = h + k - 1 - i \geqq h$ and

$$[\gamma HG^i, \gamma KG^j] \leqq [H, \gamma_{h+1}(G)] = 1.$$

If $i \geqq k$, then for a similar reason $[\gamma HG^i, \gamma KG^j] = 1$. It follows from (8) that $\gamma([H, K])G^{h+k-1} = 1$ and hence that $[H, K] \leqq \zeta_{h+k-1}(G)$.

To prove the second part observe that by (8)

$$\gamma([\gamma HG^a, \gamma HG^b])\, G^c \leqq \prod_{i+j=c} [\gamma HG^{a+i}, \gamma HG^{b+j}]. \tag{9}$$

Let $a + b + c \geqq 2h - 1$ and suppose that $i + j = c$; then

$$(a + i) + (b + j) \geqq 2h - 1,$$

which means that $a + i \geqq h$ or $b + j \geqq h$. In either case

$$[\gamma HG^{a+i}, \gamma HG^{b+j}] \leqq [\gamma_{h+1}(G), H] = 1.$$

The required result now follows from (9). ▯

Proof of Theorem 4.25. Let G be a group for which $\gamma_{i+1}(G)$ is finite and let $C = C_G(\gamma_{i+1}(G))$. If $0 \leqq s \leqq i$, we define

$$F_s = \gamma CG^{i-s}/(\gamma CG^{i-s}) \cap \zeta_{i+s}(G).$$

Since $\gamma CG^i \leqq \gamma_{i+1}(G)$, the group F_0 is finite. Suppose that F_s is finite where $s < i$: we will prove that F_{s+1} is finite.

By the second part of Lemma 4.26 we have $[\gamma CG^{i-s-1}, C] \leqq \zeta_{i+s}(G)$. Hence $\gamma CG^{i-s-1}/\gamma CG^{i-s} \cap \zeta_{i+s}(G)$ is a central factor of C. This implies that for any $g \in G$ the map $x \to [x, g]\,((\gamma CG^{i-s}) \cap \zeta_{i+s}(G))$ is a homomorphism of γCG^{i-s-1} into F_s; if $K(g)$ is the kernel of this homomorphism, $\gamma CG^{i-s-1}/K(g)$ is finite.

Now G/C is finite since $\gamma_{i+1}(G)$ is finite; let $\{g_1, \ldots, g_r\}$ be a transversal to C in G and write

$$K = K(g_1) \cap \cdots \cap K(g_r).$$

Clearly $\gamma CG^{i-s-1}/K$ is finite. Now $[K, C] \leqq [\gamma CG^{i-s-1}, C] \leqq \zeta_{i+s}(G)$ and $[K, g_j] \leqq [K(g_j), g_j] \leqq \zeta_{i+s}(G)$, by definition of K_j. Finally, $G = \langle g_1, \ldots, g_r, C\rangle$, so $[K, G] \leqq \zeta_{i+s}(G)$, which shows that

$$K \leqq (\gamma CG^{i-s-1}) \cap \zeta_{i+s+1}(G).$$

It follows that F_{s+1} is finite, as a homomorphic image of $\gamma CG^{i-s-1}/K$. By induction on s we conclude that F_i is finite and clearly

$$F_i \simeq C\zeta_{2i}(G)/\zeta_{2i}(G).$$

Since G/C is finite, so is $G/\zeta_{2i}(G)$. ▯

A closer inspection of the proof reveals that we may draw two further conclusions.

(i) If $\gamma_{i+1}(G)$ has finite order d, the index of $\zeta_{2i}(G)$ does not exceed

$$d^{(\log_2 d)^{2i} + \log_2 d}.$$

The case $i = 1$ is in a paper of Macdonald [2] (Theorem 6.1).

(ii) *If $\mathfrak{X}$ is a class of groups such that*

$$\mathfrak{X} = \boldsymbol{S}_n\mathfrak{X} = \boldsymbol{H}\mathfrak{X} = \boldsymbol{P}\mathfrak{X} = \boldsymbol{R}_0\mathfrak{X} \leqq \mathfrak{G}$$

and such that an $\mathfrak{X}\mathfrak{N}$-group of automorphisms of an $\mathfrak{X}$-group is an $\mathfrak{X}$-group, then $\gamma_{i+1}(G) \in \mathfrak{X}$ always implies that $G/\zeta_{2i}(G) \in \mathfrak{X}$.

For example, by Theorem 3.27 we could take $\mathfrak{X}$ to be the class of polycyclic groups.

Hall's Third Problem. Concise and Verbose Words

A word θ is said to be *concise* if whenever θ is finite-valued on a group G, it always follows that $\theta(G)$ is finite: a word which is not concise is called *verbose* (The terminology here is due to Turner-Smith [2]). Thus Problem (III) asks if there are any verbose words.

Let θ be a word in n variables and suppose that θ is finite-valued on a group G. Let $g, g_1, \ldots, g_n$ be elements of G. Then

$$(\theta(g_1, \ldots, g_n))^g = \theta(g_1^g, \ldots, g_n^g);$$

hence $x = \theta(g_1, \ldots, g_n)$ has only finitely many conjugates in G and $|G: C_G(x)|$ is finite. Clearly $\theta(G)$ is finitely generated, so $G/C_G(\theta(G))$ is finite. Hence $\theta(G)$ *is central-by-finite and* $\theta(G)$ *will be finite if it is periodic.*

Lemma 4.27 (P. Hall, unpublished; Turner-Smith [2]). Every verbose word is a commutator word.

Proof. Let θ be a word in $x_1, \ldots, x_n$. If θ is not a commutator word, then for some i the sum of the exponents of x_i in θ is non-zero: let this sum be r. Suppose that θ is finite-valued on a group G and let $g \in G$. If we replace x_i by g and x_j by 1 when $j \neq i$, then θ assumes the value g^r. Hence the set $\{g^r : g \in G\}$ is finite. This excludes the possibility that g have infinite order, so G is periodic and $\theta(G)$ is finite. Hence θ is concise. ▯

Thus the search for verbose words narrows down to the commutator words. Of the latter the most manageable are the lower central words; these were shown to be concise by P. Hall and also by Turner-Smith [2]. Turner-Smith has also proved that the *derived words* δ_i defined by $\delta_0 = \gamma_1$ and $\delta_{i+1} = [\delta_i, \delta_i]$ are all concise [2]; the arguments in this case are already complicated. It is unsettled whether all outer commutator words are concise.

Lemma 4.28. Let $G = \langle X \rangle$ and let $A = \langle \alpha_1, \ldots, \alpha_m \rangle \leqq \operatorname{Aut} G$. Assume that the set of all $[x, \alpha_i]$, $x \in X$, $i = 1, \ldots, m$, is finite. Then the elements of X lie in finitely many right cosets of $C_G(A)$.

Proof. Suppose that on the contrary $x_1, x_2, \ldots$ is an infinite sequence of elements of X belonging to distinct right cosets of $C = C_G(A)$. Let P be the set product of the finite sets $S_i = \{[x, \alpha_i]: x \in X\}$, where $i = 1, \ldots, m$, and let π be the mapping of X into P under which $x \to ([x, \alpha_1], \ldots, [x, \alpha_m])$. Then $(x_k)\pi = (x_l)\pi$ if and only if $x_k x_l^{-1}$ is fixed by each α_i, i.e. $Cx_k = Cx_l$. Hence the $(x_k)\pi$ are distinct and consequently P is infinite, which is a contradiction. □

Theorem 4.29 (Turner-Smith [2]). Let θ and ϕ be concise words and let $\psi = [\theta, \phi]$. Assume that ψ is finite-valued on a group G and that $\phi(\mathrm{Aut}_G\, \theta(G))$ is finitely generated. Then $\psi(G)$ is finite.

Proof. Let $C_\theta = C_G(\theta(G))$ and $C_\phi = C_G(\phi(G))$. By hypothesis $\phi(G/C_\theta) = \phi(G)C_\theta/C_\theta$ is finitely generated. Lemma 4.28 with $\theta(G)$ for G and $\phi(G)C_\theta/C_\theta$ for A is applicable, since ψ is finite-valued on G; therefore the values of θ on G lie inside finitely many cosets of C_θ, which is to say that θ is finite-valued on G/C_ϕ. Since θ is concise, $\theta(G/C_\phi) = \theta(G)C_\phi/C_\phi$ is finite. Consequently the same argument can be applied with the roles of θ and ϕ interchanged and we can conclude that $\phi(G)\,C_\theta/C_\theta$ is finite. Thus $\theta(G)/\theta(G) \cap C_\phi$ and $\phi(G)/\phi(G) \cap C_\theta$ are finite and we have the familiar situation of the Corollary to Theorem 4.13; it follows that $\psi(G) = [\theta(G), \phi(G)]$ is finite. □

Corollary. If θ is a concise word, then so is $[\theta, \gamma_1]$. Hence all the lower central words are concise.

Proof. Assume that $\psi = [\theta, \gamma_1]$ is finite-valued on a group G. Then clearly $\psi(G) = \psi(H)$ for some finitely generated subgroup H. Now $\gamma_1(\mathrm{Aut}_H\, \theta(H)) = \mathrm{Aut}_H\, \theta(H) \simeq H/C_H(\theta(H))$, which is finitely generated; hence $\psi(H)$ is finite since γ_1 is obviously concise. □

Theorem 4.29.1 (Turner-Smith [2]). If θ is a word such that $(\theta, \mathfrak{F})$ is a Schur pair, then θ is concise on the class of residually finite groups.

Proof. Let G be a residually finite group and assume that θ involves n variables and is finite-valued on G: we have to prove that $\theta(G)$ is finite. Certainly there are only finitely many elements of the form

$$\theta(g_1, \ldots, g_i a, \ldots, g_n)\, \theta(g_1, \ldots, g_i, \ldots, g_n)^{-1},$$

where $a \in G$ and $g_i \in G$. Hence we can find a normal subgroup N with finite index in G which contains no non-trivial elements of this form. Evidently N is θ-marginal and $N \leq \theta^*(G)$. Therefore $G/\theta^*(G)$ is finite, and because $(\theta, \mathfrak{F})$ is a Schur pair, $\theta(G)$ is finite. □

In view of Corollary 1 to Theorem 4.21 we can state

Corollary. Every outer commutator word is concise on the class of residually finite groups.

Some further results of a general character are known: *all three of Hall's problems have a positive solution in the class of linear groups* (Merzljakov [4]) *and in the class* $(\boldsymbol{R}\mathfrak{F})^{\boldsymbol{H}}$ (Turner-Smith [2]). Both of these classes contain, for example, all polycyclic groups (Theorem 10.11 and Corollary 2, Theorem 9.31) and the second class contains all finitely generated abelian-by-nilpotent groups (Theorem 9.51).

4.3 Groups with Finite Conjugacy Classes

If G is a group and x is an element of G, then x is called an *FC-element* of G if x has only a finite number of conjugates in G, that is, if $|G: C_G(x)|$ is finite. *This is also equivalent to* $|G: C_G(x^G)|$ *being finite*: for $C_G(x^G) \leqq C_G(x)$ and if x has n conjugates in G, the elements of G permute these by conjugation and consequently $|G: C_G(x^G)|$ has finite order dividing $n!$.

Lemma 4.31. In any group G the FC-elements form a characteristic subgroup F and $G/C_G(F)$ is residually finite (Baer [13], § 2, Proposition 1).

Proof. Let x and y be FC-elements of G; then both $C_G(x)$ and $C_G(y)$ have finite index in G, so $C_G(x) \cap C_G(y)$ also has finite index. But $C_G(x) \cap C_G(y) \leqq C_G(xy^{-1})$, so $|G: C_G(xy^{-1})|$ is finite and xy^{-1} is an FC-element. Hence the set F of all FC-elements is a characteristic subgroup. Since $C_G(F)$ is the intersection of all the $C_G(x)$ where $x \in F$, the group $G/C_G(F)$ is residually finite. $\square$

The subgroup of all FC-elements of a group is called the *FC-centre.* A group is an *FC-group* if it coincides with its FC-centre, i.e., if every conjugacy class is finite.

FC-groups were first studied by Baer [12] and B. H. Neumann [7], who obtained the basic properties; see also Erdős [1] and Černikov [24] for alternative treatments. Numerous papers on FC-groups have been published since this initial period.

The following result on the structure of FC-groups is basic.

Theorem 4.32. Let G be an FC-group.

(i) $G/\zeta(G)$ is locally normal and finite (Baer [12], § 2, Theorem 2).

(ii) The elements of finite order in G form a subgroup containing G' which is locally normal and finite (B. H. Neumann [7]).

Proof. Let $x \in G$ and let $\{t_1, \ldots, t_k\}$ be a right transversal to $C_G(x)$ in G. Each $C_G(t_i^G)$ is normal and has finite index in G. Hence there is a positive integer $m = m(x)$ such that $x^m \in C_G(t_i^G)$ for $i = 1, \ldots, k$. Thus x^m centralizes $t_1, \ldots, t_k$ and also, of course, $C_G(x)$, which implies that $x^m \in \zeta(G)$ and $G/\zeta(G)$ is periodic. Since the class of FC-groups is clearly

H-closed, $G/\zeta(G)$ is a periodic FC-group. Now in any periodic FC-group the conjugates of a given element form a normal finite subset; hence, by Dietzmann's Lemma (Corollary 2 to Lemma 2.14), they generate a normal finite subgroup. Thus *every periodic FC-group is locally normal and finite.* In particular $G/\zeta(G)$ is locally finite and the Corollary to Theorem 4.12 implies that G' is locally finite. It follows that the set of elements of finite order form a subgroup containing G' and, by the obvious **S**-closure of the class of FC-groups, this subgroup is locally normal and finite. □

Corollary 1. A group is an FC-group if and only if it can be embedded in the direct product of a periodic FC-group and a torsion-free abelian group.

Proof. The class of FC-groups is easily seen to be **D**-closed as well as **S**-closed, so the condition is sufficient. Conversely, let G be an FC-group and let T be the subgroup of elements of finite order in G: then G/T is a torsion-free abelian group since $G' \leqq T$. By Zorn's Lemma there is a maximal torsion-free subgroup of $\zeta(G)$, say X; clearly $\zeta(G)/X$ is periodic and $X \lhd G$, so G/X is periodic. Since $X \cap T = 1$, the mapping $g \to (gX, gT)$ is an embedding of G into $(G/X) \times (G/T)$. □

Corollary 2. (B. H. Neumann [7]). A finitely generated group is an FC-group if and only if it is a finite extension of a finitely generated subgroup of its centre.

For a finitely generated FC-group is central-by-finite, either by Theorem 4.32 or from first principles. Conversely, every central-by-finite group is an FC-group.

Corollary 3 (Polovickiĭ [7]). A group G is an FC-group if and only if x^G is finite-by-cyclic for every $x \in G$.

Proof. Let G be an FC-group and let $x \in G$. Then x^G is a finitely generated FC-group, so its elements of finite order form a finite subgroup, by Corollary 2. Now $[G, x] \leqq G'$, which is periodic by Theorem 4.32, so $[G, x]$ is finite; thus $x^G = \langle x \rangle [G, x]$ is finite-by-cyclic since $[G, x) \lhd G$. Conversely, it is easy to show that if $H \in \mathfrak{F}\mathfrak{C}$, then Aut H is finite; thus if $x^G \in \mathfrak{F}\mathfrak{C}$, then $G/C_G(x^G)$ is finite. □

Černikov [24] has pointed out that a *periodic FC-extension of a central torsion-free subgroup is an FC-group*. This partial converse of Theorem 4.32 is easily proved by noting that the elements of finite order form a subgroup with abelian factor group and applying the method of proof of Corollary 1, Theorem 4.32 above. A similar result is in Baer's paper [12] (§ 2, Theorem 2).

An interesting consequence of the basic theorem on FC-groups is a solution to what might be called the dual of Schmidt's Problem.

Theorem 4.33 (Fedorov [1]; see also Erdős [1], Theorem 7.1). An infinite group has each non-trivial subgroup of finite index if and only if it is infinite cyclic.

Proof. Let G be an infinite group in which every non-trivial subgroup has finite index. Let $1 \neq x \in G$; then $|G: \langle x \rangle|$ is finite, so G is finitely generated and x has infinite order; thus G is torsion-free. Also G is an FC-group since $\langle x \rangle \leq C_G(x)$. Theorem 4.32 shows at once that G is free abelian of finite rank and it is clear from this that G must be infinite cyclic. The converse is obvious. ▯

Corollary (B. H. Neumann [11], Theorem 6.3). A group has a finite covering of cyclic subgroups if and only if it is either infinite cyclic or finite.

Proof. Let G be covered by a finite set of cyclic subgroups; by Lemma 4.17 we can assume that all of these have finite index in G. Let $1 \neq g \in G$; then g belongs to a cyclic subgroup C such $|G: C|$ is finite. Since $|C: \langle g \rangle|$ must be finite, $|G: \langle g \rangle|$ is finite and Fedorov's theorem implies that G is either infinite cyclic or finite. The converse is clearly true. ▯

In a similar way it may be shown that *a group which has a finite covering of FC-subgroups is an FC-group* (B. H. Neumann [11], p. 241).

We remark that in contrast with Theorem 4.33 infinite groups which have all their proper factor groups finite form a wide class, containing for example all infinite simple groups. McCarthy [1], [2] calls such groups *just-infinite* and studies soluble just-infinite groups: see also Theorem 5.41 below.

Periodic FC-Groups

The first corollary to Theorem 4.32 has the effect of focussing attention on the periodic FC-groups, which we know to be just the locally normal and finite groups. Obviously, any direct product of finite groups and all its subgroups are periodic FC-groups and are also residually finite. Conversely, *one may ask whether residually finite, periodic FC-groups can be embedded in a direct product of finite groups.* As yet no complete solution to this question has been obtained but for countable groups the answer is known to be positive.

Theorem 4.34 (P. Hall [8], p. 290). The countable, residually finite, periodic FC-groups are precisely the groups which can be embedded in direct products of countably many finite groups.

Proof. Let G be a countable, residually finite, periodic FC-group. Then G is locally normal and finite, so there is a countable ascending chain of normal finite subgroups whose union is G, say

$$1 = G_0 \leqq G_1 \leqq \cdots.$$

We will show how to construct a descending chain of normal subgroups

$$G = R_1 \geqq R_2 \geqq \cdots$$

such that G/R_i is finite and $G_i \cap R_i = 1$ for each i. Suppose that $R_1, \ldots, R_i$ have already been chosen. Since G is residually finite and G_{i+1} is finite, there is a normal subgroup K with finite index in G such that $G_{i+1} \cap K = 1$. Let $R_{i+1} = K \cap R_i$: then G/R_{i+1} is finite and

$$G_{i+1} \cap R_{i+1} = (G_{i+1} \cap K) \cap R_i = 1.$$

Thus our construction has been effected.

Now we define

$$S_{i+1} = G_i R_{i+1}, \quad (i = 0, 1, 2, \ldots).$$

Certainly $S_{i+1} \lhd G$ and G/S_{i+1} is finite. In addition

$$G_{i+1} \cap S_{i+1} = G_{i+1} \cap (G_i R_{i+1}) = G_i(G_{i+1} \cap R_{i+1}) = G_i.$$

If $1 \neq g \in G$, then $g \in G_{i+1} \backslash G_i$ for some i and therefore $g \notin S_{i+1}$. Hence

$$\bigcap_{i=1,2,\ldots} S_i = 1.$$

Now an element of G belongs to all but a finite number of the G_i and therefore to all but a finite number of S_i. Hence the monomorphism $g \to (gS_1, gS_2, \ldots)$ of G into the cartesian product of the G/S_i is actually into the direct product of the G/S_i. □

Since every group of order n can be embedded in the symmetric group S_n of degree n, we deduce that there exists a universal countable, residually finite, periodic FC-group

Corollary. The countable, residually finite, periodic FC-groups are up to isomorphism just the subgroups of the group $S_2 \times S_3 \times \cdots \times S_n \times \cdots$

We mention in passing that P. Hall [9] has constructed a universal countable locally finite group which is simple.

If G is an FC-group, $G/\zeta(G)$ is residually finite, by Lemma 4.31, and by Theorem 4.32 it is also periodic. Hence the following is a special case of Hall's theorem: *a countable FC-group with trivial centre can be embedded in a direct product of finite groups*. These results of P. Hall have been extended by Gorčakov [1] as follows: *a residually finite, periodic FC-group can be embedded in a direct product of finite groups if* *its centre is countable*: see also Kargapolov [2] and Gorčakov [2] and [5] for special cases and other results on this problem. Another sufficient condition for a periodic FC-group to be embeddable in a direct product of finite groups is that each Sylow subgroup be finite (see (ii), p. 139); this is due to Černikov [25].

On the other hand, in [8] P. Hall has constructed an example of an uncountable FC-group of exponent 4 with derived subgroup of order 2 which is not even a homomorphic image of a subgroup of a direct product of finite groups. Hall also proves that *every countable periodic FC-group is a homomorphic image of a subgroup of some direct product of finite groups*. Thus countability would appear to play an essential role here.

Further Results on *FC*-Groups and Related Topics

(i) The class of FC-groups is not $\mathbf{L}$, $\mathbf{R}$, $\mathbf{P}$ *or* $\mathbf{N}_0$*-closed.*
It is easy to find locally finite groups and residually finite groups which are not FC-groups; this disposes of $\mathbf{L}$ and $\mathbf{R}$-closure. Since $\mathbf{N}_0 \leqq \mathbf{PH}$, it is sufficient to prove that the class of FC-groups is not $\mathbf{N}_0$-closed. In fact the product of two normal abelian subgroups may very easily fail to be an FC-group. For example, let A and B be two abelian groups and let $F = A * B$ be their free product. Then $F' = [A, B]$ and $N = [A, B, B]\,[B; A, A] \lhd F$. The group

$$G = F/N$$

is the *second nilpotent product of A and B* (see Schenkman [7], p. 178). It is easy to show that the mapping $a \otimes b \rightarrow [a, b]\,N$ is an isomorphism of $A \otimes B$ with $G' = [A, B]/N$ (Wiegold [2], Theorem 3.10, MacHenry [1]). Since $[G', G] = 1$, the group G is the product of two normal abelian subgroups, $A[A, B]/N$ and $B[A, B]/N$. If we choose A and B so that for some fixed $a \in A$ the set of all $a \otimes b$, $b \in B$, is infinite, then G is not a FC-group: for example, let A have order 2 and let B be an infinite elementary abelian 2-group.

(ii) The number of elements in each conjugacy class of a group G is a power of a fixed prime p if and only if G can be embedded in the direct product of a p-group which is also an FC-group and an abelian group without elements of order p. In particular such a group is a hypercentral group with hypercentral length at most ω (cf. Macdonald [1]).

Suppose that G satisfies the condition on conjugacy classes. By the argument of Corollary 1 to Theorem 4.32 we may confine ourselves to the case where G is periodic. Let x be an arbitrary element of G; then x^G is finite. If $|x^G|$ is prime to p, then $x^G \leqq C_G(y)$ for every $y \in G$ and consequently $x \in \zeta(G)$. On the other hand, if $|x^G|$ is divisible by p, then, by writing x^G as a union of G-conjugacy classes, we see that at least one of these conjugacy classes apart from $\{1\}$ has a single element only; thus $x^G \cap \zeta(G) \neq 1$. This argument shows that $x \in \zeta_n(G)$ where $n = |x^G|$. Hence $G = \zeta_\omega(G)$ and G is hypercentral of length at most ω. Thus G is the direct product of its p and p'-components, the latter being abelian.

On the other hand, it is clear that if D is a direct product of the type in question, the number of elements in a conjugacy class of D is a power of p. Indeed the same is true of any subgroup of D: for if $C \leq D$ and if $|D: C|$ is a power of p, the hypercentrality of D implies that there is an $N \lhd G$ such $N \leq C$ and D/N is a finite p-group.

(iii) Groups with boundedly finite conjugacy classes

A group G is called a *BFC-group* if each conjugacy class is finite and the number of its elements does not exceed some number $d = d(G)$. B. H. Neumann has characterized BFC-groups in the following manner.

Theorem 4.35 (B. H. Neumann [9], Theorem 3.1). A group G is a BFC-group if and only if G' is finite.

Proof. If G' has finite order d and $x \in G$, the number of commutators $[x, y]$, $y \in G$, is at most d; hence the number of conjugates x^y cannot exceed d.

Conversely let G be a BFC-group; suppose that d is the maximum number of elements in a conjugacy class of G and let a be an element of G with exactly d conjugates. Let $\{b_1, \ldots, b_d\}$ be a right transversal to $C_G(a)$ in G and observe that $a^{b_1}, \ldots, a^{b_d}$ are the distinct conjugates of a. Define

$$C = \bigcap_{i=1}^{d} C_G(b_i),$$

so that $|G: C|$ is finite; let $\{c_1, \ldots, c_k\}$ be a right transversal to C in G. Now we write

$$N = \langle a, c_1, \ldots, c_k \rangle^G$$

and observe that N is finitely generated since each generator has only a finite number of conjugates, and $G = NC$. Let $x \in C$, then $(xa)^{b_i} = xa^{b_i}$, which shows that the d conjugates $(xa)^{b_i}$ are all distinct and therefore exhaust the conjugates of xa. Now let $y \in C$: then $(xa)^y = xa^{b_i}$ for some i and consequently $[x, y] = x^{-1}x^y = a^{b_i}(a^y)^{-1} \in N$. It follows that $C' \leq N$ and since $G = NC$, this implies that $G' \leq NC' \leq N$. By Theorem 4.32 the group G' is periodic and by Corollary 2 to Theorem 4.32 the elements of finite order in the finitely generated FC-group N form a finite subgroup. Therefore G' is finite. ☐

The proof can be made to yield a fairly crude bound for $|G'|$ in terms of $d = d(G)$, the maximum number of elements in a conjugacy class of G, when this is finite. Improved bounds for $|G'|$ have been given by Wiegold [1], Macdonald [2] and Shepperd and Wiegold [1]. It has been conjectured by Wiegold [1] that the inequality

$$|G'| \leq d^{\frac{1}{2}(1+\log_2 d)}$$

holds, but it seems to be difficult to decide this question. It is not hard to reduce the problem to the case where G is a finite group (see Macdonald [2], Lemma 2.1). Recently Bride [1] has confirmed Wiegold's conjecture for nilpotent groups of class at most 2. For other recent work on this problem see P. M. Neumann [2] and Wiegold [8].

(iv) Further results of B. H. Neumann

In his paper [12] B. H. Neumann has shown that *the groups with finite derived subgroup can also be characterized as those groups G such that $|H^G : H|$ is finite for every subgroup H of G*; see also Scott [3]. In the same paper Neumann proved that *the central-by-finite groups are precisely the groups in which every subgroup has only a finite number of conjugates* (cf. Theorem 4.47 below). Macdonald [2] has obtained upper bounds for the order of the derived subgroup and the index of the centre and his paper contains a full discussion of these problems.

The second result of Neumann mentioned above was subsequently generalized by Eremin [1] as follows: *a group is central-by-finite if and only if each abelian subgroup has only a finite number of conjugates.* In [2] and [4] Eremin discusses groups in which every infinite subgroup has finitely many conjugates.

(v) In [1] Schiefelbusch studies the *Trofimov number* of a group G: this is the greatest common divisor of the finite indices of normalizers of non-normal subgroups of G, or 0 if there are no such indices; he shows that *if G is either FC or residually finite, the Trofimov number of G is* 0, 1 *or a prime.* The Trofimov number of finite groups has been studied by Trofimov [1] and Hering [1].

(vi) Groups with Černikov conjugacy classes

Polovickiĭ [7] has considered groups with Černikov conjugacy classes or *CC-groups*: these are the groups G such that $G/C_G(x^G)$ is a Černikov group for all x in G. In view of the second definition of an FC-element (p. 121) *every FC-group is a CC-group.* Concerning CC-groups we shall prove

Theorem 4.36 (cf. Polovickiĭ [7]). The following statements about a group G are equivalent.

(i) $[G, x]$ is a Černikov group and $G/C_G([G, x])$ is periodic for every $x \in G$.

(ii) x^G is Černikov-by-cyclic and $G/C_G(x^G)$ is periodic for every $x \in G$.

(iii) G is a CC-group.

Proof.

(i) implies (ii). Let G satisfy (i). Clearly $x^G/[G, x]$ is cyclic and therefore has finite automorphism group; also x^G is Černikov-by-cyclic. G induces in $[G, x]$ a periodic group of automorphisms and from this it is easy to see that G does the same in x^G.

(ii) implies (iii). According to Theorem 3.29, a periodic group of automorphisms of a Černikov group is itself a Černikov group. This implies that a periodic group of automorphisms of a Černikov-by-cyclic group is also a Černikov group.

(iii) implies (i). We assume that G is a CC-group. If X is a finite subset of G, then

$$C_G(X^G) = \bigcap_{x \in X} C_G(x^G);$$

consequently $G/C_G(X^G)$ is a Černikov group by the obvious $\boldsymbol{R}_0$-closure of the class of Černikov groups. Hence X^G is an extension of its centre by a Černikov group and, by Theorem 4.23, the group $(X^G)'$ is Černikov. Thus, in particular, G' is locally finite.

From now on x will be a fixed element of G and A will denote x^G. Since A' is a Černikov group, we can assume that A is abelian. Let $C = C_G(A)$, so that $A \leqq C$, and let R/C be the finite residual of the Černikov group G/C. Then G/R is finite and R/C is a radicable abelian group satisfying Min. Let $r \in R$; then $(\langle x, r\rangle^G)'$ is a Černikov group, by the first paragraph of the proof. Hence $[r, A]$ is a Černikov group. Since $R' \leqq C = C_G(A)$,

$$[r, A]^R \leqq [r^R, A] = [r, A],$$

and $[r, A] \lhd R$. Now $[r, A] \leqq A$, so $C_R([r, A]) \geqq C$ and $\bar{R} = R/C_R([r, A])$ is a periodic radicable group. By the Corollary to Theorem 3.29.2 the group $\bar{R}$ is finite—here we use the commutativity of A. Since $\bar{R}$ is radicable, it must be trivial and $[r, A]$ lies in the centre of R. Hence

$$[R, A, R] = 1. \tag{10}$$

Since $[C, x] = 1$, the mapping $Cr \rightarrow [r, x]$ is well-defined, and by (10) it is a homomorphism of R/C onto $[R, x]$. Consequently $[R, x]$ is a Černikov group.

Now choose a transversal $\{t_1, \ldots, t_n\}$ to R in G. Evidently

$$[G, x] = \langle [R, x]^{t_i}, [t_i, x] : i = 1, \ldots, n\rangle.$$

Each $[R, x]^{t_i}$ is a normal Černikov subgroup of R, so the product L of all these subgroups is a normal Černikov subgroup of G (we are using the $\boldsymbol{P}$ and $\boldsymbol{H}$-closure of the class of Černikov groups). Also the elements

$[t_i, x]$ generate a finite subgroup F because G' is locally finite. Finally, $[G, x] = LF$, so $[G, x]$ is a Černikov group while $C = C_G(x^G) \leq C_G([G, x])$ and $G/C_G([G, x])$ is periodic, as required. □

Corollary (Polovickiĭ [7]). If G is a CC-group, G' is locally normal and Černikov and, in particular, is locally finite.

Groups which Have Finitely Many Conjugacy Classes

In contrast to FC-groups, groups which have only a finite number of conjugacy classes can have a very complicated structure. For, *if G is a group which has elements of just s distinct finite orders, G can be embedded in a group with exactly n conjugacy classes for each integer n such that $n > s + 1$.* (Gorčinskiĭ [1], Theorem 1). In the case $s = 0$, when G is torsion-free, this theorem is due to Higman, Neumann and Neumann [1]. Some further results on this subject are in another paper of Gorčinskiĭ [2].

On the other hand, Burnside's theorem that an irreducible group of linear operators with degree n over an algebraically closed field contains n^2 linearly independent elements may be used to show that *a linear group over a field is finite if it has only a finite number of conjugacy classes* (see Curtis and Reiner [1], p. 262, Exercise 1): for linear groups over the complex number field this was first proved by Burnside [5].

FC-Nilpotent, *FC*-Hypercentral and *FC*-Soluble Groups

A normal series $\{\Lambda_\sigma, V_\sigma : \sigma \in \Sigma\}$ in a group G is said to be *FC-central* if Λ_σ/V_σ lies in the FC-centre of G/V_σ for each $\sigma \in \Sigma$. A group which possesses an FC-central series of finite length is called *FC-nilpotent* (Haimo [2], p. 498), and a group which possesses an ascending FC-series is called *FC-hypercentral*; these are, of course, generalizations of nilpotence and hypercentrality.

In any group G one can form *the upper FC-central series* $\{F_\alpha\}$ defined by the rules

$$F_0 = 1,\ F_{\alpha+1}/F_\alpha = \text{the } FC\text{-centre of } G/F_\alpha \text{ and } F_\lambda = \bigcup_{\beta<\lambda} F_\beta,$$

where α is an ordinal and λ is a limit ordinal. This a special case of the upper χ-central series defined in Section 1.3. The limit of the upper FC-central series is the *FC-hypercentre.* Clearly $\zeta_\alpha(G) \leq F_\alpha(G)$ is always true: G is FC-nilpotent if and only if $G = F_c$ for some finite c and G is FC-hypercentral if and only if G coincides with its FC-hypercentre.

Similarly a group is called *FC-soluble* if it is a poly-FC-group (Duguid and McLain [1]); this is a wider class than the FC-nilpotent groups and includes all soluble groups.

Many properties of nilpotent, hypercentral and soluble groups can be carried over to FC-nilpotent, FC-hypercentral and FC-soluble groups without undue difficulty. For example, *the classes of FC-nilpotent and FC-hypercentral groups are* $\mathbf{N}_0$*-closed*, as a routine adaptation of the proof of the $\mathbf{N}_0$-closure of the class of hypercentral groups shows (p. 51). The class of FC-soluble groups is $\mathbf{N}_0$-closed by $\boldsymbol{P}$ and $\boldsymbol{H}$-closure.

Hypercentre and *FC*-Hypercentre

McLain has investigated relations between the hypercentre and the FC-hypercentre of a group. In the next two theorems we shall write F_α and Z_α respectively for the αth terms of the upper FC-central and the upper central series of the specified group G.

Theorem 4.37 (McLain [4]). If G is a group with torsion-free centre and if H denotes the hypercentre of G, then $F_\alpha \cap H = Z_\alpha$ for each ordinal α.

Proof. Since $Z_\alpha \leq F_\alpha$ is universally valid, $Z_\alpha \leq F_\alpha \cap H$. Suppose that α is the first ordinal for which $F_\alpha \cap H \nleq Z_\alpha$ and let $x \in (F_\alpha \cap H)\backslash Z_\alpha$. Clearly α is not a limit ordinal, so there is an element g such that $[x, g] \notin Z_{\alpha-1}$. Since x belongs to H, so does $[x, g]$ and $[x, g] \in Z_{\beta+1}\backslash Z_\beta$ where $\alpha - 1 < \beta + 1$, i.e. $\alpha - 1 \leq \beta$. Now $x \in F_\alpha$ and therefore $|G: C_G(xF_{\alpha-1})|$ is finite, from which it follows that $g^m \in C_G(xF_{\alpha-1})$ for some positive integer m. Hence $[x, g^m] \in F_{\alpha-1} \cap H = Z_{\alpha-1}$, by minimality of α. Now

$$[x, g]^m \equiv [x, g^m] \equiv 1 \bmod Z_\beta$$

since $Z_{\alpha-1} \leq Z_\beta$, and by Theorem 2.25 the group $Z_{\beta+1}/Z_\beta$ is torsion-free. Consequently $[x, g] \in Z_\beta$, which is impossible. □

Theorem 4.38 (McLain [4]). If G is a locally nilpotent group, then

$$Z_\alpha \leq F_\alpha \leq Z_{\omega\alpha}$$

for each ordinal α.

Proof. Of course it is obvious that $Z_\alpha \leq F_\alpha$. Suppose that $F_\beta \leq Z_{\omega\beta}$ holds for all $\beta < \alpha$. If α is a limit ordinal, it follows immediately that $F_\alpha \leq Z_{\omega\alpha}$. If α is not a limit ordinal, $F_{\alpha-1} \leq Z_{\omega(\alpha-1)}$ and $F_\alpha Z_{\omega(\alpha-1)}/Z_{\omega(\alpha-1)}$ lies in the FC-centre of $G/Z_{\omega(\alpha-1)}$; since $\zeta_\omega(G/Z_{\omega(\alpha-1)}) = Z_{\omega\alpha}/Z_{\omega(\alpha-1)}$, it is sufficient to prove the result when $\alpha = 1$.

Let $x \in F_1$ and put $N = x^G$. Then N is finitely generated and $G/C_G(N)$ is finite. Let T be a transversal to $C_G(N)$ in G and set $H = \langle N, T\rangle$, a finitely generated and therefore nilpotent subgroup. We shall prove that

$$N \cap \zeta_i(H) \leq Z_i$$

for each integer $i \geq 0$. For $i = 0$ this obvious, so let $i > 0$ and assume that $N \cap \zeta_{i-1}(H) \leq Z_{i-1}$. Now $N \cap \zeta_i(H)$ is centralized by $C_G(N)$ and

$[N \cap \zeta_i(H), T] \leqq N \cap \zeta_{i-1}(H) \leqq Z_{i-1}$. Therefore $[N \cap \zeta_i(H), G] \leqq Z_{i-1}$ and consequently $N \cap \zeta_i(H) \leqq Z_i$ as required.

Now let the nilpotent class of H be c. Then $N = N \cap \zeta_c(H) \leqq Z_c$, so $x \in Z_c$ and therefore $F_1 \leqq Z_\omega$. □

Corollary 1. In a locally nilpotent group the hypercentre and the *FC*-hypercentre coincide.

Corollary 2. In a locally nilpotent group with torsion-free centre the corresponding terms of the upper central and upper *FC*-central series coincide.

The first corollary follows at once from Theorem 4.38. If G satisfies the hypotheses of the second corollary, then F_α lies in H, the hypercentre of G, by Corollary 1, and $F_\alpha = F_\alpha \cap H = Z_\alpha$ by Theorem 4.37.

Since there exist torsion-free hypercentral groups of arbitrary hypercentral length (Gluškov [5], McLain [4]), Corollary 2 shows that *there exist FC-hypercentral groups of arbitrary FC-hypercentral length*: see also Haimo [2] and Duguid [1] on this question.

Imposition of Finiteness Conditions

We consider now the effect of imposing the conditions Max-*ab* or Min-*ab* on *FC*-soluble, *FC*-nilpotent and *FC*-hypercentral groups. First of all we deal with the case of *FC*-groups.

Theorem 4.39. (i) An *FC*-group satisfies Min-*ab* if and only if it is a central-by-finite Černikov group (Baer [23], p. 521).

(ii) An *FC*-group satisfies Max-*ab* if and only if it is a finitely generated central-by-finite group.

Proof. Let G be an *FC*-group and assume that G satisfies Min-*ab*. By Lemma 4.31 the finite residual R of G lies in the centre of G, so R satisfies Min. Hence G/R satisfies Min-*ab* (see p. 88). Thus we can assume that G is residually finite and prove that G is finite. Suppose that G is infinite. Assume that $x_1, \ldots, x_n$ are non-trivial elements of G such that the subgroup they generate is the direct product of $\langle x_1 \rangle, \ldots, \langle x_n \rangle$. Since G is periodic, each x_i^G is finite and hence $P = x_1^G \ldots x_n^G$ is finite. Now G is residually finite, so there is a normal subgroup N with finite index in G such that $P \cap N = 1$; also $N \neq 1$ because G is infinite. Now choose a non-trivial element x_{n+1} from N and observe that $P \cap x_{n+1}^G = 1$. Thus $[P, x_{n+1}^G] = 1$ and $\langle x_1, \ldots, x_{n+1} \rangle$ is the direct product of $\langle x_1 \rangle, \ldots, \langle x_{n+1} \rangle$. However this construction leads to a subgroup of G which is the direct product of infinitely many non-trivial cyclic groups and this contradicts Min-*ab*.

Now let G be an FC-group which satisfies Max-*ab*. Then $\zeta(G)$ is finitely generated; also $G/\zeta(G)$ satisfies Max-*ab* (see p. 86) and by Theorem 4.32 it is periodic. Hence $G/\zeta(G)$ satisfies Min-*ab* and is a Černikov group, by (i). Finally $G/\zeta(G)$ is finite by the property Max-*ab*.

In each case the converse assertion is obviously true. $\square$

Theorem 4.39.1. An FC-soluble group satisfies Max-*ab* (Min-*ab*) if and only if it is a polycyclic-by-finite (Černikov) group.

This was first proved by Baer ([23], p. 530) in the case of the property Min-*ab*. For a generalization see Section 5.4, p. 180.

Proof. Let G be an FC-soluble group which satisfies Max-*ab* and let $1 = G_0 \lhd G_1 \lhd \cdots \lhd G_n = G$ be an FC-series of G with finite length. By Theorem 3.31 the group G has a unique largest normal radical subgroup R which is polycyclic and contains all subnormal soluble subgroups of G. Now G/R satisfies Max-*ab* and clearly it contains no nontrivial subnormal abelian subgroups. Thus we may assume that $R = 1$. It now follows from Theorem 4.39 that G_1 is finite; G_2/G_1 is an FC-group which satisfies Max-*ab*, so it is central-by-finite, by Theorem 4.39 again. Hence there exists H *sn* G such that $G_1 \leqq H \lhd G_2$, H/G_1 is abelian and G_2/H is finite. Now an elementary argument (or alternatively Theorem 4.25) shows that

$$\mathfrak{F}\mathfrak{A} \leqq \mathfrak{N}_2\mathfrak{F}.$$

Hence H is finite since G has no subnormal abelian subgroups except 1. It follows that G_2 is finite and by induction on n we find that G is finite. The converse assertion is obvious.

The proof for Min-*ab* is exactly analogous. $\square$

In particular Theorem 4.39.1 describes those FC-soluble groups that satisfy Max or Min: here the result is due to Duguid and McLain [1]; see also Nishigôri [2].

Theorem 4.39.2. An FC-hypercentral group satisfies Max-*ab* (Min-*ab*) if and only if it is a finitely generated nilpotent-by-finite group (a Černikov group).

Proof. In both cases the sufficiency of the conditions is easy to establish. In the case of Min-*ab* the necessity follows directly from Theorem 3.36. Suppose that G is an FC-hypercentral group with Max-*ab*. Then the Corollary to Theorem 3.36 shows that G satisfies Max. Hence G has an FC-central series of finite length, say $1 = G_0 \leqq G_1 \leqq \cdots \leqq G_n = G$. By Max the factor G_{i+1}/G_i is finitely generated, so G induces in it a finite group of automorphisms. Let N be the intersection of all the $C_G(G_{i+1}/G_i)$: then G/N is finite and N is nilpotent. $\square$

Observe that Theorem 4.39.1 and 4.39.2 show that *for groups which satisfy Min-ab the properties FC-nilpotent, FC-hypercentral and FC-soluble are identical* and *for groups which satisfy Max-ab, FC-nilpotent and FC-hypercentral are the same*; however, even for groups which satisfy Max, FC-soluble is a weaker property. For example, the polycyclic group

$$\langle x, a, b \colon [a, b] = 1, a^x = b, b^x = ab \rangle$$

has trivial FC-centre.

In addition there is the result of McLain [4] that *a finitely generated FC-hypercentral group is nilpotent-by-finite,* which implies that FC-hypercentral groups are locally FC-nilpotent. The proof parallels that of the local nilpotency of hypercentral groups (p. 50).

The papers of Haimo [2], Nishigôri [2] and Pic [2] contain other results of an elementary character about FC-soluble and FC-nilpotent groups.

4.4 The Classification of Groups with Finite or Černikov Layers

Let G be a group and let m be a positive integer or infinity; then we call the subgroup G_m generated by all elements of G with order m the *m-layer* of G. When m is finite, G_m may be regarded as a dual of the mth power $G^m = \langle g^m \colon g \in G \rangle$. It is clear that G is covered by the normal subgroups G_m.

All the elements of a conjugacy class of G have the same order, so they belong to the same layer of G; thus imposition of restrictions on the layers of a group will affect the conjugacy classes. The principal aim of this section is the classification of groups all of whose layers are either finite or Černikov groups. A group whose layers are finite is clearly a rather special type of FC-group.

As evidence of the considerable influence on a group of its layers, we present first a simpler result.

Theorem 4.41 (Polovickiĭ [6]). A group whose layers are abelian is itself abelian.

Proof. Let G be a group with abelian layers. For any $x \in G$ the subgroup x^G lies within a layer, so it is abelian and G is the product of normal abelian subgroups. By Fitting's theorem on the N_0-closure of the class $\mathfrak{N}$ (Theorem 2.18), G is locally nilpotent, and from this it follows that the elements of finite order form a subgroup of G. If G has an element of infinite order, then G can be generated by elements of infinite order, i.e. G coincides with its ∞-layer. Hence we may assume that G is periodic and, since G is locally nilpotent and therefore the direct product of its Sylow subgroups, we can even assume that G is a p-group.

Suppose that G is not abelian and let x be an element of least order that does not belong to the centre of G; let y be an element of least order that does not commute with x. Then if x, y and xy have orders p^l, p^m and p^n respectively, it follows that $1 \leqq l < m < n$; for no two of x, y and xy commute. Now $x^G y^G$ is nilpotent of class 2 by Theorem 2.18; consequently the well-known formula

$$(xy)^{p^m} = x^{p^m} y^{p^m} [y, x]^{\binom{p^m}{2}} = [y, x]^{\binom{p^m}{2}}$$

is valid. Now $m \geqq 2$, so p divides $\binom{p^m}{2}$; also $[y, x, x] = 1$. Hence

$$(xy)^{p^m} = [y, x^p]^{\frac{1}{2}p^{m-1}(p^m-1)} = 1,$$

since x^p belongs to the centre of G. This implies that $n \leqq m$, which is not the case. □

Corollary. A group has cyclic layers if and only if it is infinite cyclic or a subgroup of the multiplicative group of all complex roots of unity.

A related result is the paper [3] of Dlab; dual problems for powers of a group are treated in a cycle of papers by Szász [1]—[4], [6] and Dlab [1], [2].

CL-Groups and *FL*-Groups

A group each of whose layers is a Černikov group is called a *CL-group* and a group whose layers are all finite is called an *FL-group*. Obviously every *FL*-group is a *CL*-group. Since every group is the product of its layers and since Černikov groups are countable and locally finite, *CL-groups are countable and locally finite.*

FL-groups can be characterized as those groups which have just a finite number of elements of each order, including ∞; for *FL*-groups clearly have the latter property and if G is a group with this property, it can have no elements of infinite order (otherwise it would have infinitely many of them) and the elements of order m form a normal finite subset; consequently the m-layer which they generate is finite by Dietzmann's Lemma.

FL-groups were first studied by Baer [12] and Černikov [12] in 1948, while *CL*-groups were introduced by Polovickiĭ [1] in 1961.

Polovickiĭ's Classification of *CL*- and *FL*-Groups

In a series of papers Polovickiĭ showed that the *CL*-groups and *FL*-groups are exhausted by the subgroups of certain direct products of Černikov groups and also characterized these groups in terms of their normal and Sylow structure.

We will call a direct product of groups ***prime-thin*** if for each prime p at most a finite number of the direct factors contain elements of order p; this terminology is due to Černikov [25].

Theorem 4.42 (Polovickiĭ [2], [4]). The following properties of a group G are identical.

(i) G is a CL-group.

(ii) G is locally normal and Černikov and each Sylow subgroup satisfies Min-*ab*.

(iii) G is isomorphic with a subgroup of a prime-thin direct product of Černikov groups.

Theorem 4.43 (Polovickiĭ [2], [4]). The following properties of a group G are identical.

(i) G is an FL-group.

(ii) G is locally normal and finite and each Sylow subgroup satisfies Min-*ab*.

(iii) G is isomorphic with a subgroup of a prime-thin direct product of central-by-finite Černikov groups.

Polovickiĭ proved these results in the course of his investigation of the π-minimal condition (see p. 74), the equivalence of (i) and (ii) in Theorem 4.43 having been previously established by Černikov [25]. The proofs presented here are essentially those in the paper [14] of Robinson.

We begin with three elementary lemmas.

Lemma 4.44 (Fuchs [3], p. 68, Exercise 20). An abelian group is a CL-group if and only if it is a direct product of cyclic p-groups and groups of type p^∞ with only a finite number of such direct factors for each prime p.

This follows from well-known facts about abelian groups: it implies that for abelian groups the properties CL and FL are identical: more generally, this is true for nilpotent groups since a nilpotent group which is generated by elements of order m has exponent dividing a power of m, by Theorem 2.26.

Lemma 4.45. If G is an FL-group, Aut G is residually finite.

Proof. Let $1 \neq \alpha \in \operatorname{Aut} G$; then $g^\alpha \neq g$ for some $g \in G$. Denote by A the subgroup of all automorphisms which fix g; then $\alpha \notin A$ and A has finite index in Aut G since the g^β, $\beta \in \operatorname{Aut} A$, have the same order as g and consequently their number is finite. □

Lemma 4.46 (cf. Černikov [26], § 2, Roseblade [3], Robinson [4], p. 36). The class of periodic radicable abelian groups is $\acute{\boldsymbol{N}}$-closed.

Proof. By Lemma 1.31 it will be enough to establish $\boldsymbol{N}_0$-closure. Let H and K be two periodic radicable abelian subgroups of a group G and assume that $H \lhd G$ and $K \lhd G$. Then $[H, K, K] \leqq K' = 1$ and Lemma 3.13 implies that $[H, K] = 1$. Hence HK is abelian and clearly it is also periodic and radicable. ▯

Corollary. If H *asc* G and H is a periodic radicable abelian group, then $H \lhd^2 G$.

We leave the simple proof of the corollary to the reader.

Proof of Theorem 4.42. (a) It is clear that a subgroup of a prime-thin direct product of Černikov groups is a CL-group, so certainly (iii) implies (i). Let G be a CL-group. If $x \in G$, then x^G lies inside a layer, so it is a Černikov group. Hence, if $\{x_1, \ldots, x_n\}$ is any finite subset,

$$\{x_1, \ldots, x_n\} \subseteq x_1^G \ldots x_n^G,$$

which is a Černikov group by the Corollary to Theorem 3.12. Thus G is locally normal and Černikov. Each elementary abelian p-subgroup of G is finite, so the Sylow p-subgroups of G satisfy Min-*ab*. Thus (i) implies (ii) and it remains to show that (ii) implies (iii).

(b) Let G be a group satisfying condition (ii); then G is an extension of a periodic radicable abelian CL-group by a group which contains only finitely many elements of order a power of p for each prime p.
Let Q be a quasicyclic subgroup of G and let X be any finite subset of G. Then X^G is a Černikov group, so it has a characteristic radicable abelian subgroup A which satisfies Min and has finite index. By Lemma 4.45 the group Aut A is residually finite; hence Q centralizes A as well as X^G/A. Thus $[X^G, Q, Q] = 1$ and it follows that $[G, Q, Q] = 1$. Therefore Q *sn* G. Let R be the subgroup generated by all the quasicyclic subgroups of G. Then Lemma 4.46 shows that R is a periodic radicable abelian group; also each primary component of R satisfies Min; therefore R is a CL-group.

Let p be any prime and let P/R be a Sylow p-subgroup of G/R. Since G is locally finite and its Sylow p-subgroups satisfy Min-*ab*, the Corollary to Lemma 3.46 shows that P/R satisfies Min-*ab*; thus, by Theorem 3.32, it is a Černikov group. Let F/R be the finite residual of P/R; then P/F is finite and F/R is a radicable abelian group satisfying Min. Let q be any prime and denote by R_q the q'-component of R. Since F/R_q is a soluble group without proper subgroups of finite index and also satisfies Min, Theorem 3.12 shows that it is abelian. Thus $F' \leqq R_q$ for all q, so that F is abelian and radicable and $F = R$. It follows that P/R is finite; hence each Sylow subgroup of G/R is finite.

Now let P_1/R be any other Sylow p-subgroup of G/R. Then P/R and P_1/R together generate a finite group of which they are also Sylow p-subgroups; thus P/R and P_1/R are conjugate. But P/R lies in a normal Černikov subgroup of G/R which must even be finite since all Sylow subgroups of G/R are finite. Thus P_1/R also lies in this normal finite subgroup and it follows at once that G/R can contain only a finite number of elements with order a power of p.

(c) For each prime p the elements with order a power of p generate a Černikov subgroup of G.
Let R be as in (b) and let $\{x_1R, \dots, x_nR\}$ be the set of elements with order a power of p in G/R. Then $X = \langle x_1, \dots, x_n \rangle$ is a finite group and hence X^R is a Černikov group. The p-component of R, say P, is also a Černikov group, so PX^R is a Černikov group. We will prove that each element of G with order a power of p belongs to PX^R. If x has order p^m, then $x = x_i r$ where $1 \leqq i \leqq n$ and $r \in R$. Hence

$$1 = (x_i r)^{p^m} \equiv x_i^{p^m} r^{p^m} \bmod [X, R].$$

Consequently $r^{p^m} \in X[X, R] = X^R$, so $r \in PX^R$ and $x = x_i r \in PX^R$.

(d) Every homomorphic image of G satisfies condition (ii) *of Theorem 4.42.*
For if H is a homomorphic image of G, it is obvious that H is a locally normal and Černikov group and the Corollary to Lemma 3.46 shows that the Sylow subgroups of H satisfy Min-*ab*.

(e) If p is a prime and M is the maximal normal p'-subgroup of G, then G/M is a Černikov group.
Let $H = G/M$, so that H satisfies condition (ii) and has no non-trivial normal p'-subgroups. Let S be the subgroup generated by all quasicyclic subgroups of H; then S is abelian by *(b)*, so it must be a p-group; hence it satisfies Min. By *(c)* the elements of order a power of p generate a normal finite subgroup of H/S, say T/S. Then T is a Černikov group; let $C = C_H(T)$ and note that H/C is a Černikov group by Theorem 3.29. Now the elements of C which have order a power of p lie in $T \cap C \leqq \zeta(C)$, so that $C/\zeta(C)$ is a locally finite p'-group. By the Corollary to Theorem 4.12, the elements of C with order prime to p form a characteristic subgroup N. But $N \lhd H$; hence $N = 1$ and C is a p-group and therefore a Černikov group. It follows that H is a Černikov group.

(f) The final step.
Let $p_1, p_2, \dots,$ be the primes which divide orders of elements of G and let M_i be the maximal normal p_i'-subgroup of G. By (e) each G/M_i is a Černikov group and it is clear that the intersection of all the M_i is trivial. Hence the map $g \to (gM_1, gM_2, \dots)$ is an embedding of G in the cartesian

product of the G/M_i. Let p be any prime and let P be the subgroup generated by all elements of G with order a power of p. By *(c)* the group P is Černikov and the set π of all primes dividing orders of elements of P is finite. If $p_i \notin \pi$, then $P \leq M_i$ and G/M_i is a p'-group. Therefore only a finite number of the G/M_i contain elements of order p. It follows that the elements of finite order in the cartesian product of the G/M_i are just the elements of the direct product. Hence G is embedded in the direct product of the G/M_i and this direct product is, of course, prime-thin. □

Proof of Theorem 4.43. Let G be an FL-group; then if $x \in G$, the subgroup x^G is finite since it lies inside a layer. Thus G is locally normal and finite; clearly each Sylow subgroup satisfies Min-*ab*. Hence (i) implies (ii). Let G satisfy (ii). By Theorem 4.42 the group G can be embedded in a prime-thin direct product of Černikov groups, each of which can clearly be taken to be a homomorphic image of G. But a homomorphic image of G which satisfies Min, being an FC-group, is central-by-finite, by Lemma 4.31. Hence (ii) implies (iii).

Finally, to prove that (iii) implies (i) it is sufficient to show that *each central-by-finite Černikov group H is an FL-group.* Let m be a positive integer, let $Z = \zeta(H)$ and let $\{t_1, \ldots, t_n\}$ be a fixed transversal to Z in G. If x is an element of G with order m, then $x = t_i z$ where $1 \leq i \leq n$ and $z \in Z$. Denote by l the least common multiple of m and the orders of the t_i; then $1 = x^l = t_i^l z^l = z^l$. Since Z satisfies Min, the possibilities for z—and hence for x—are finite in number. Therefore G is an FL-group. □

Some Further Properties of CL-Groups and FL-Groups

(i) The class of FL-groups and the class of CL-groups are **H**-*closed* (Baer [12], p. 1030 and Černikov [12], Theorem 2).
These statements follow from Theorems 4.42 and 4.43 and the Corollary to Theorem 3.46.

(ii) In a CL-group the finite residual is radicable and abelian and is generated by all the quasicyclic subgroups (Robinson [14]).
For let G be a CL-group, let F be the finite residual of G and let R be the subgroup generated by all the quasicyclic subgroups of G; then R is radicable and abelian and $R \leq F$. Now G/R is residually a Černikov group, by Theorem 4.42, and any homomorphic image of G/R which is a Černikov group is actually finite, because G/R has only finitely many elements of order a power of a given prime p (by part (b) of the proof of Theorem 4.42). Hence G/R is residually finite and $R = F$.

A group whose Sylow subgroups are finite is said to be *thin* (Černikov [12]): thus for CL-groups "thin" and "residually finite" are identical properties. Clearly a thin CL-group is an FL-group and *a group is a thin*

CL-group if and only if for each prime p it contains just a finite number of elements of order a power p. Hence the class of thin *CL*-groups is **H**-closed. *The thin CL-groups can also be characterized as the locally normal and finite groups whose Sylow subgroups are finite, and as the subgroups of prime-thin direct products of finite groups* (Černikov [23], [25]). These results follow easily from Theorem 4.42.

(iii) A CL-group is the product of its finite residual and a thin CL-group (Polovickiĭ [4]).

This was proved by Eremin for *FL*-groups in [3]. The result may be derived from Theorem 4.42: we merely sketch the proof. Let G be a *CL*-group with finite residual R and suppose that G is embedded as a subgroup G_1 in the prime-thin direct product D of the groups $G/M(p)$; here $M(p)$ is the maximal normal p'-subgroup of G. Observe that the result is true for D and also that the finite residual of $G/M(p)$ is just $RM(p)/M(p)$, which is isomorphic with the p-component of R. Now in the embedding R is mapped onto a subgroup R_1 of the direct product of the groups $RM(p)/M(p)$. Since R is isomorphic with this direct product and since each primary component of R satisfies Min, it follows that R_1 equals the direct product and coincides with the finite residual of D. Now $D = R_1X$ where X is thin and $R_1 \leqq G_1$, so

$$G_1 = G_1 \cap (R_1X) = R_1(G_1 \cap X)$$

and of course $G_1 \cap X$ is thin.

(iv) The class of CL-groups and the class of thin CL-groups are ascendant coalition classes, but neither is a subnormal coalition-class (Robinson [14], Theorem 4).

(v) We mention yet another characterization of central-by-finite Černikov groups. *Each non-trivial homomorphic image of a group has at least one but only finitely many elements of square-free order* >1 *if only if it is a central-by-finite Černikov group* (Baer [23], p. 521).

For if G is a group with the first property, it is easily seen to be the union of a chain of normal finite subgroups and to satisfy Min-*ab*. Thus G is an *FC*-group and by Theorem 4.39 it is a central-by-finite Černikov group. On the other hand, it was shown during the proof of Theorem 4.43 that a group of the latter type is an *FL*-group, and it is clear that the number of square-free orders that can arise is finite.

(vi) Further properties of *CL* and *FL*-groups are to be found in Berlinkov [1], Černikov [12], [23], [25], Polovickiĭ [1]—[4] and Robinson [14]. We remark that in the last paper it is proved that *a group whose layers are polycyclic-by-finite is either an FL-group or a polycyclic-by-finite group* (Theorem 3). In [2] Polovickiĭ studies the problem of embedding locally normal and Černikov groups in direct products of Černikov

groups and obtains an analogue of the theorem of P. Hall on periodic FC-groups (p. 123).

An Indecomposable CL-Group

Černikov ([12], § 4) has constructed an infinite, directly indecomposable, thin FL-group; this shows that *a CL-group need not in general be a direct product of Černikov groups*, but only a subgroup of such a direct product. We will describe Černikov's construction.

First of all a suitable sequence of primes $p_1, p_2, \ldots$ is chosen. Let p_1 be an arbitrary prime and suppose that $p_1, \ldots, p_{2i-1}$ have already been chosen. Let p_{2i+1} be any prime not already selected and let p_{2i} be a prime such that

$$p_{2i} \equiv 1 \bmod p_{2i-1}p_{2i+1}.$$

The existence of p_{2i} is guaranteed by Dirichlet's theorem on the infinity of primes in an arithmetic progression (Hasse [2], p. 167).

Let $\langle a_i \rangle$ be a cyclic group of order p_i and let

$$A = \langle a_1 \rangle \times \langle a_3 \rangle \times \cdots \quad \text{and} \quad B = \langle a_2 \rangle \times \langle a_4 \rangle \times \cdots.$$

If $i > 0$, then p_{2i+1} divides $p_{2i} - 1$ and there exists a primitive p_{2i+1}th root of unity λ_i in the field $GF(p_{2i})$; similarily if $i \geq 0$, there is a primitive p_{2i+1}th root of unity μ_i in $GF(p_{2i+2})$. We make A into a group of automorphisms of B by defining

$$a_{2i}^{a_{2i+1}} = a_{2i}^{\lambda_i}, \quad a_{2i+2}^{a_{2i+1}} = a_{2i+2}^{\mu_i} \tag{11}$$

and in all other cases

$$[a_r, a_s] = 1. \tag{12}$$

Let G be the semi-direct product of B by A; then G is a periodic metabelian group in which each Sylow subgroup is cyclic of prime order. It is clear that G has just a finite number of elements of order p_{2i}. Suppose that g is an element of G with order p_{2i+1}: then $g = ba_{2i+1}^r$ where $b \in B$ and $0 < r < p_{2i+1}$. Now $[b, a_{2i+1}] \in [B, a_{2i+1}] = \langle a_{2i} \rangle \times \langle a_{2i+2} \rangle$ if $i > 0$, by (11) and (12). Hence

$$1 = g^{p_{2i+1}} \equiv b^{p_{2i+1}} \bmod \langle a_{2i} \rangle \times \langle a_{2i+2} \rangle.$$

b has order prime to p_{2i+1}, so $b \in \langle a_{2i} \rangle \times \langle a_{2i+1} \rangle$ and $g \in \langle a_{2i}, a_{2i+1}, a_{2i+2} \rangle$, which is a finite group. When $i = 0$, the same argument leads to $g \in \langle a_1, a_2 \rangle$. Hence G contains finitely many elements of order p_{2i+1}, and therefore G is a thin FL-group.

Finally, suppose that $G = H \times K$ and let $a_i = hk$ where $h \in H$ and $k \in K$; then both h and k have order 1 or p_i. If $h \neq 1$ and $k \neq 1$, then

$\langle h\rangle\times\langle k\rangle$ has order p_i^2, which is impossible because the Sylow subgroups of G are cyclic. Hence $h = 1$ or $k = 1$ and a_i belongs to H or to K. Now $[a_i, a_{i+1}] \neq 1$ by (11); thus a_i and a_{i+1} lie in the same direct factor. This implies that $G = H$ or K and G is indecomposable.

Groups with Finite Classes of Isomorphic Subgroups

Theorem 4.47 (Fuchs [1]). A group G has finite classes of isomorphic subgroups if and only if it is a finite extension of a subgroup of its centre which is isomorphic with a subgroup of K, the multiplicative group of complex roots of unity.

Proof. Assume that the classes of isomorphic subgroups of G are finite; this means that the number of subgroups of G isomorphic with a given subgroup is finite. If x is an element of G with infinite order, the groups $\langle x\rangle, \langle x^2\rangle, \langle x^3\rangle, \ldots$ are isomorphic. But this is impossible, so G is periodic. Next, if $\{x_1, x_2, \ldots\}$ is an infinite set of elements with the same finite order, all the elements x_i must lie in the union of finitely many subgroups $\langle x_1\rangle, \langle x_2\rangle, \ldots, \langle x_k\rangle$. By this contradiction G is an FL-group. In particular, G is locally normal and finite, each Sylow subgroup of G is a Černikov group and each quasicyclic subgroup of G lies in Z, the centre of G.

Let π be the set of all primes p for which there exist at least two distinct cyclic p-subgroups with the same order (say $p^{m(p)}$). Observe that if $p \notin \pi$, there is a unique Sylow p-subgroup of G and this is locally cyclic, i.e. it is finite cyclic or of type p^∞.

Suppose that the set π is infinite. We will construct an infinite sequence of distinct primes in π, say $p_1, p_2, \ldots$, such that if P_i is the subgroup generated by all the elements of order $p_i^{m(p_i)}$, then $[P_i, P_j] = 1$ if $i \neq j$. Let p_1 be any prime in π and assume that $p_1, \ldots, p_i$ have already been chosen. Clearly each P_j is finite, so $L = P_1 \cdots P_i$ is finite; also of course $L \lhd G$. Let p_{i+1} be a prime in π which has not already been chosen and which does not divide $|\mathrm{Aut}\, L|$. Then $P_{i+1} \leq C_G(L)$ and our construction is effected. Let a_i be an element of order $p_i^{m(p_i)}$; then the elements $a_1, a_2, \ldots$ generate the direct product $X = \langle a_1\rangle\times\langle a_2\rangle\times\cdots$. But each $\langle a_i\rangle$ can be chosen in at least two ways, so there are infinitely many distinct choices for X, all of them isomorphic. By this contradiction π is finite.

Let S be the subgroup generated by Z and the Sylow p-subgroups for $p \notin \pi$. Then by definition of π we see that S is a normal abelian subgroup of G and G/S contains elements of only finitely many distinct prime orders. Now all the quasicyclic subgroups of G lie in Z and hence in S: thus in G/S the elements with order a power of a prime p form a finite set. (Here we use step (b) in the proof of Theorem 4.42.) It follows that G/S is finite. Let $\{t_1, \ldots, t_m\}$ be a transversal to S in G. Each t_i^G is finite and normal in G and hence is centralized by all but a finite number of the Sylow subgroups of G. But an element which lies in S and centralizes each t_i^G belongs to Z, the centre of G; hence S/Z is finite and consequently G/Z is finite. Finally, it is clear that G cannot have a subgroup which is the direct product of two groups of type p^∞ and, since all but a finite number of the Sylow subgroups of G are cyclic or quasicyclic, Z has a subgroup of finite index which is isomorphic with a subgroup of K.

Conversely, let G be a group with a subgroup N such that $N \leqq \zeta(G)$, $|G:N|$ is finite and N is isomorphic with a subgroup of K. There is a finite subgroup X such that $G = XN$. Let $N = L \times M$ where L is the product of all the Sylow p-subgroups of N for p dividing $|X|$. Then $G = (XL)M$ and $(XL) \cap M = (XL) \cap N \cap M = L \cap M = 1$ since $X \cap N \leqq L$. Also $X \lhd G$, $L \lhd G$ and $M \lhd G$, so $G = H \times M$ where $H = XL$. Clearly H and M have no elements of the same prime order and H contains elements of only finitely many distinct prime orders. Let S and T be isomorphic subgroups of G. Then $S = (S \cap H) \times (S \cap M)$ and $S \simeq T$ is equivalent to $S \cap H \simeq T \cap H$ and $S \cap M \simeq T \cap M$; moreover the latter isomorphism implies that $S \cap M = T \cap M$ in view of the structure of N. Hence we can assume that $G = H$ and G contains elements of only finitely many distinct prime orders. Therefore G satisfies Min.

Let S_1 and T_1 be the finite residuals of S and T; then S/S_1 and T/T_1 are finite and S_1 and T_1 are contained in N. Clearly $S_1 \simeq T_1$, so we must have $S_1 = T_1$. It follows that we may assume that S is finite, say of order d. Then S and all subgroups isomorphic to it lie in the union of the i-layers of G for i dividing d; but this is finite since G is an FL-group, being a central-by-finite Černikov group. This completes the proof. □

It will be observed that in the first part of the proof it is only necessary to assume that the isomorphism classes of *abelian* soubgroups of G are finite.

Chapter 5

Finiteness Conditions on the Normal and Subnormal Structure of a Group

In this chapter we are concerned with finiteness conditions which refer only to the normal or subnormal subgroups of a group. Now a simple group has only two subnormal subgroups and its subnormal structure is therefore very obvious; yet a simple group may have exceedingly complex general structure—as we will show in the first section of this chapter. For this reason we will often investigate finiteness conditions referring to the normal or subnormal structure of a group in conjunction with an additional property such as generalized solubility of a suitable type.

5.1 Infinite Simple Groups. Embedding Theorems

The easiest example of an infinite simple group is probably the alternating group (or group of even finitary permutations) on an infinite set; this is simple because the union of any chain of simple groups is simple.

The existence of finitely generated, infinite simple groups was first demonstrated by Higman [2] in the following manner: let G be the group with generators

$$a, b, c, d$$

and defining relations

$$a^b = a^2,\ b^c = b^2,\ c^d = c^2,\ d^a = d^2, \tag{1}$$

and let N be a maximal normal subgroup of G; such an N exists by Zorn's Lemma. $H = G/N$ is a finitely generated simple group which must be infinite, because the relations (1) are incompatible with finite orders for a, b, c and d: for details see Higman [2].

Subsequently, Camm [1] constructed explicitly $2^{\aleph_0}$ non-isomorphic 2-generator simple groups. Calenko [1] showed that *there are* $2^{\mathfrak{c}}$ *non-isomorphic simple groups with cardinality* $\mathfrak{c}$ *for each infinite cardinal* $\mathfrak{c}$.

The work of Kostrikin [4] and of Novikov and Adjan [1] shows that *there exist finitely generated infinite simple groups of prime exponent* p *for each* $p \geqq 4381$. For, let B be the infinite Burnside group $B(n, p)$ where $n > 1$ and $p \geqq 4381$. By a theorem of Kostrikin there is a maximal finite factor group B/N of B. Clearly N is finitely generated and non-trivial and N has no proper subgroups with finite index. Let M be a maximal normal subgroup of N. Then N/M is a group of the required kind.

Further examples of infinite simple groups may be found in Chehata [1], Clowes and Hirsch [1], P. Hall [13] and Higman [4]; see also Section 8.4.

Embedding Groups in Simple Groups

It has been known for some time that *every group can be embedded in a simple group*. This may be deduced from a theorem of Baer on the normal structure of the symmetric group on an arbitrary infinite set X (Baer [3]; see also Karrass and Solitar [1]); if S is the full symmetric group on X and if N is the subgroup of all elements s of S such that the set of elements moved by s has cardinality less than that of X, then $N \lhd S$ and S/N is a simple group. Let G be any group, which we can assume to be infinite, and let ϱ be the right regular representation of G; then taking $X = G$ in Baer's result and noting that $G^\varrho \cap N = 1$, we obtain $G \simeq G^\varrho N/N \leqq S/N$.

It is also well-known that *every finite group can be embedded in a finite simple group*; for if G is a finite group of order $n > 2$, we first embed G in the symmetric group of degree n and then embed the latter in the alternating group of degree $n + 2$ by adjoining an additional 2-cycle to odd permutations.

Theorem 5.11. Let H be an arbitrary group. Then H can be embedded as a subgroup of a simple group H^* in such a way that every isomorphism between two finitely generated subgroups of H can be extended to an inner automorphism of H^*. Moreover, if H is infinite, H^* can be taken to have the same cardinality as H.

A special case of this theorem has been found by Calenko [1]. To prove the theorem we use three lemmas which are of some interest in themselves.

Lemma 5.12. Let G and G^* be two permutation groups on a set X and assume that both are semi-regular (i.e. only the identity fixes an element of X). Suppose that there is a one-one mapping of the set of G-orbits onto the set of G^*-orbits. Then, if $g \to g^*$ is an isomorphism of G with G^*, there is a permutation π of X such that $g^* = \pi^{-1} g \pi$ for all $g \in G$.

Proof. From each G-orbit select one element and denote the resulting set of elements by T; for G^* we do likewise and obtain a set T^*. By hypothesis there is a one-one mapping $t \to t^*$ of T onto T^*. Now G is semi-regular, so each element of X can be written *uniquely* in the form tg

where $t \in T$ and $g \in G$. Since G^* is also semi-regular, the mapping $tg \to t^*g^*$ defines a permutation π of X. Let $h \in G$; then

$$(t^*g^*)\,\pi^{-1}h\pi = (tg)\,h\pi = t^*(gh)^* = (t^*g^*)\,h^*.$$

Hence $\pi^{-1}h\pi = h^*$. ☐

Lemma 5.13 (Higman, Neumann and Neumann [1]) Let H be an arbitrary group. Then H can be embedded in a group G in such a way that every isomorphism between two subgroups of H can be extended to an inner automorphism of G.

Proof. Let K and K^* be two isomorphic subgroups of H and let $x \to x^*$ be an isomorphism between K and K^*. If H is infinite, it is possible for $|H:K|$ and $|H:K^*|$ to be different and we first rectify this. Let H be embedded as a direct factor in $\bar{H} = H \times H_1$ where $H_1 \simeq H$. If H has infinite cardinality $\mathfrak{c}$, then

$$\mathfrak{c} = |H| \leqq |\bar{H}:K| \leqq |\bar{H}| = \mathfrak{c}^2 = \mathfrak{c}.$$

Hence $|\bar{H}:K| = \mathfrak{c}$ and in the same way $|\bar{H}:K^*| = \mathfrak{c}$. Thus $|\bar{H}:K| = |\bar{H}:K^*|$; of course this is also true if H is finite.

Now we embed $\bar{H}$ in S, the full symmetric group on $\bar{H}$, by means of the right regular representation, so that K and K^* become isomorphic semi-regular permutation groups on $\bar{H}$. The set of K-orbits has cardinality $|\bar{H}:K|$ and the set of K^*-orbits has cardinality $|\bar{H}:K^*|$. Hence there exists a π in S such that $x^* = \pi^{-1}x\pi$ for all $x \in K$ by Lemma 5.12. Finally, let G be the subgroup of S generated by $\bar{H}$ and all such π. ☐

Lemma 5.14. Let H and G be arbitrary groups. If G is non-trivial, H can be embedded in a group which is generated by isomorphic copies of G.

Proof. First we form the standard wreath product $K = H \wr T$ where $T = \langle t \rangle$ is a group of order 2. The base group of K is the direct square $H \times H$ of H. Denote by D the diagonal subgroup consisting of all (h, h) with $h \in H$; clearly $D \simeq H$. If $h \in H$,

$$(t(h, 1))^2 = t^2(1, h)\,(h, 1) = (h, h).$$

Hence we can identity H with D and conclude that each element of H is the square of some element of K.

Next we embed K in a group L in which all elements of K with the same order are conjugate; this is possible by Lemma 5.13. Let $h \in H$. If h has odd or infinite order, then $h^x = h^2$ for some $x \in L$ and $h = [h, x] \in L'$. If h has order a power of 2, then $h = k^2$ and $k^x = k^3$ for some $k \in K$ and $x \in L$, and $h = k^2 = [k, x] \in L'$. Hence $H \leqq L'$.

Finally, let $M = L \wr G$, the standard wreath product of L by G. We identify L with the direct factor L_1 of the base group. Let $a, b \in L$ and let $1 \neq g \in G$; then a and b^g belong to different direct factors L_1 and L_g of the base group, so they commute. Hence

$$1 = [a, b^g] = [a, b\,[b, g]] = [a, [b, g]]\,[a, b]^{[b,g]},$$

which shows that $[a, b] \in G^M$. Thus $H \leq L' \leq G^M$ and G^M is generated by copies of G. ∎

Proof of Theorem 5.11. Let H be an arbitrary group. We first embed H in a group H_1 which contains elements of all possible orders: for example we can take H_1 to be the direct product of H with the multiplicative group of complex roots of unity and an infinite cyclic group. Then, using Lemma 5.14, we embed H_1 in a group H_2 which can be generated by involutions.

Next let H_3 be the standard wreath product $T \wr H_2$ where T is a group of order 2; if B denotes the base group of the wreath product, then H_3 is a semi-direct product of B by H_2 and clearly $C_{H_3}(B) = B$. Thus a non-trivial normal subgroup of H_3 must intersect B non-trivially.

We define a sequence of groups $H_3, H_4, \ldots$, with H_n embedded in H_{n+1} for $n = 3, 4, \ldots$, such that all isomorphisms between finitely generated subgroups of H_n extend to inner automorphisms of H_{n+1}; that such a construction is possible follows from Lemma 5.13. Let H^* be the union of the chain $H_1 \leq H_2 \leq \cdots$. Observe that these groups H_n can be chosen so that $|H^*| = |H|$ in case H is infinite: for we need adjoin at most $|H_n|$ elements to H_n to obtain H_{n+1}. It is clear that isomorphisms between finitely generated subgroups of H^* extend to inner automorphisms of H^*. Thus it remains only to show that H^* is simple.

Let $1 \neq N \lhd H^*$ and let $1 \neq x \in N$. By construction H_1 contains an element y with the same order as x and x and y are conjugate; thus $y \in N$ and $N \cap H_3 \neq 1$. Hence $N \cap B \neq 1$ and N contains an involution. Now H_2 is generated by involutions and any two such elements are conjugate; therefore $H_2 \leq N$. In particular, $H_1 \leq N$ and by construction H_1 contains elements of all possible orders. Since elements with equal orders are conjugate, $N = H^*$ and H^* is simple. ∎

In conclusion we mention an interesting unpublished theorem of P. Hall: *every countable group can be embedded in a simple group with nine generators.*

5.2 The Minimal Condition on Normal Subgroups

The minimal condition on normal subgroups is denoted by

$$\text{Min-}n.$$

The first systematic study of this property was the paper [13] of Baer, which appeared in 1949. We recall that Min-n is $\boldsymbol{P}$, $\boldsymbol{H}$ and $\boldsymbol{R}_0$-closed (Corollary, Lemma 1.48), but, of course, it is not $\boldsymbol{S}$-closed and in fact it is not even $\boldsymbol{S}_n$-closed, by an example to be constructed later in this section (p. 153). However, recently Wilson has obtained the useful fact that Min-n is inherited by subgroups with finite index.

Theorem 5.21 (Wilson [1]). If a group G satisfies Min-n, then every subgroup with finite index in G satisfies Min-n.

Proof. Suppose that H is a subgroup with finite index in G which does not satisfy Min-n. Since $G/\mathrm{Core}_G\, H$ is finite, we may assume that $H \lhd G$. Now H does not satisfy Min-H, the minimal condition on H-admissible subgroups, so, in view of the property Min-n, H contains a normal subgroup K of G which is minimal with respect to not satisfying Min-H.

Let $\mathscr{S}$ be the set of all non-empty finite subsets X of G with the following property: if

$$K_1 > K_2 > \cdots \tag{2}$$

is an infinite, strictly descending sequence of H-admissible subgroups of K, then

$$K = K_i^X \tag{3}$$

for all i. Let T be a transversal to H in G; then $G = HT$. For any chain (2) the relations $K_i \lhd H \lhd G$ imply that $K_i^T \lhd G$. Also $K_i^T \leqq K$ since $K \lhd G$; if $K_i^T < K$, then K_i^T satisfies Min-H, by minimality of K, and therefore $K_j = K_{j+1} =$ etc. for some $j \geqq i$. By this contradiction $K_i^T = K$ for all i and all chains $K_1 > K_2 > \cdots$. Thus $T \in \mathscr{S}$ and $\mathscr{S}$ is not empty.

We now select a minimal element X of $\mathscr{S}$. If $x \in X$, then $Xx^{-1} \in \mathscr{S}$ because $K \lhd G$; therefore Xx^{-1} is a minimal element of $\mathscr{S}$ containing 1. Hence we may asume that $1 \in X$. Now if $X = \{1\}$, equation (3) shows that K cannot contain an infinite strictly descending chain of H-admissible subgroups, i.e. K satisfies Min-H. Therefore X has at least two elements; let

$$X^* = X \backslash \{1\},$$

a non-empty finite subset of G which does not belong to $\mathscr{S}$, by minimality of X.

Let $K_1 > K_2 > \cdots$ be an infinite, strictly descending sequence of H-admissible subgroups of K and define

$$L_i = K_i \cap K_i^{X^*}.$$

$K_i^g \lhd H^g = H$ for all $g \in G$, so $L_i \lhd H$. Also $L_i \geqq L_{i+1}$; suppose that $L_i = L_{i+1}$. Since $X \in \mathscr{S}$, we must have $K = K_{i+1}^X$ and

$$K_i = K_i \cap K_{i+1}^X = K_i \cap (K_{i+1} K_{i+1}^{X^*}) \leqq K_{i+1} L_i = K_{i+1}.$$

Thus $K_i = K_{i+1}$, which is not the case. Hence $L_i > L_{i+1}$ and

$$L_1 > L_2 > \cdots$$

is an infinite, strictly descending sequence of H-admissible subgroups of K. Consequently $L_i^X = K$ and

$$K_i = K_i \cap L_i^X = K_i \cap (L_i L_i^{X^*}) \leq L_i(K_i \cap K_i^{X^*}) = L_i,$$

which shows that $K_i = L_i$. By definition of L_i it follows that

$$K_i^{X^*} = K_i^X = K$$

for all i, and $X^* \in \mathcal{S}$, which is a contradiction. ▯

Corollary 1 (Baer [13], p. 14, Theorem 4). If G is a group which satisfies Min-n, the centre and the FC-hypercentre of the finite residual of G coincide and satisfy Min.

Proof. Let R be the finite residual of G; since G has Min-n, the group G/R is finite and R has no proper subgroups of finite index. Let F_1 be the FC-centre of R and let F_2/F_1 be the FC-centre of R/F_1. If $x \in F_1$, then $|R: C_R(x)|$ is finite and $x \in \zeta(R)$. Hence $F_1 = \zeta(R)$ and by the same argument $F_2 = \zeta_2(R)$. Theorem 5.21 shows that R satisfies Min-n; therefore $\zeta(R)$ satisfies Min. Let $y \in R$ and $x \in \zeta_2(R)$; then $x^y = xz$, where $z \in \zeta(R)$, and the order of z divides that of x. Since $\zeta(R)$ has Min, the element x can have only finitely many conjugates in R, so $x \in F_1 = \zeta(R)$ and $F_2 = \zeta_2(R) = \zeta(R)$. It follows that $\zeta(R)$ is the FC-hypercentre of R. ▯

Corollary 2. Let χ denote the following subgroup theoretical property: $H\chi G$ if and only if $H \lhd G$ and $\mathrm{Aut}_G H$ is residually finite. Then a hyper-χ group which satisfies Min-n is a Černikov group and the properties Min and Min-n coincide for hyper-χ groups.

Proof. Let G be a hyper-χ group with the property Min-n and let R be the finite residual of G. Now G has an ascending normal χ-series and if F is any factor of this series, $\mathrm{Aut}_R F \leq \mathrm{Aut}_G F$ and consequently $\mathrm{Aut}_R F$ is resdiually finite. However, R has no proper subgroups of finite index, so it must centralize F. By intersecting R with the terms of the ascending normal χ-series we obtain an ascending central series in R. Corollary 1 now implies that R is abelian and satisfies Min, so G is a Cernikov group. ▯

If F is the FC-centre of a group G, then $G/C_G(F)$ is always residually finite, by Lemma 4.31. Hence we obtain from Corollary 2 the following theorem of Duguid and McLain [1]. *The FC-hypercentral groups which satisfy Min-n are precisely the Černikov groups and they have FC-hypercentral length* ≤ 2.

Corollary 2 also implies that *a hyperfinite group satisfying Min-n is a Černikov group* (Baer [13], p. 16, Theorem 3).

The Hypercentre and *FC*-Hypercentre in Groups Satisfying Min-*n*

Theorem 5.22 (Baer [13], pp. 11 and 20). Let G be a group which satisfies Min-n and let F_α and Z_α denote respectively the αth terms of the upper FC-central series and the upper central series of G.

(i) The FC-hypercentre of G is F_2 and it is also the unique maximal normal Černikov subgroup (and hence the Černikov radical) of G.

(ii) $F_2/\zeta(F_1)$ is finite and F_1 is the finite radical of G.

(iii) The hypercentre of G coincides with $Z_{\omega+n}$ for some finite $n \geqq 0$: in addition every upper central factor of G except perhaps Z_1 is finite.

Proof. Let R be the finite residual of G. Then $F_1 \cap R$ is contained in—and indeed coincides with—the FC-centre of R, so by Corollary 1 to Theorem 5.21 it follows that $F_1 \cap R \leqq \zeta(R)$ and $F_1 \cap R$ satisfies Min. Since $F_1/F_1 \cap R$ is finite, F_1 is a Černikov group. This argument shows also that F_2/F_1 and F_3/F_2 are Černikov groups; hence F_3 is a Černikov group, by the **P**-closure of the class of Černikov groups (Corollary to Theorem 3.12). Let S be the finite residual of F_3; then F_3/S is finite and S is a normal abelian subgroup of G with the property Min. Thus S is a union of normal finite subgroups of G and consequently $S \leqq F_1$. Hence F_3/F_1 is finite and $F_2 = F_3 =$ the FC-hypercentre of G. Clearly, every normal Černikov subgroup lies in the FC-hypercentre in any group; hence F_2 is also the unique maximal normal Černikov subgroup of G.

$F_1/\zeta(F_1)$ is finite since F_1 is an FC-group satisfying Min, and F_2/F_1 is finite, so $F_2/\zeta(F_1)$ is finite. Finally, Dietzmann's Lemma (p. 45) implies that if the FC-centre of a group is periodic, it coincides with the product of all the normal finite subgroups, i.e. with the finite radical. Thus (i) and (ii) are proved.

Now let H be the hypercentre of G. Then $H \leq F_2$ by (i) and H is a Černikov group; let T denote its finite residual. Then H/T is finite and T is abelian and satisfies Min. But a G-admissible subgroup of H with finite order m lies in Z_m (this follows from Lemma 2.16). Hence $T \leqq Z_\omega$ and H/Z_ω is finite; therefore $H = Z_{\omega+n}$ for some finite n.

Finally, we have to prove that Z_{i+1}/Z_i is finite for each finite $i > 0$. The subgroup Z_{i+1} is a Černikov group; let U be its finite residual. Then $[U, \underbrace{G, \ldots, G}_{i+1}] = 1$, while G/G' is periodic in view of Min-n. Thus $[U, G] = 1$ and $U \leqq Z_1$, by Lemma 3.13. Therefore Z_{i+1}/Z_i is finite. □

The Socle of a Group

Following Remak ([1], p. 1) we define the *socle* of a group G to be the product of all the minimal normal subgroups of G, with the understanding that should G prove to have no minimal normal subgroups the socle of G is 1.

Lemma 5.23. The subgroup generated by a set of minimal normal subgroups of a group is the direct product of certain members of that set.

Proof. Let $\{M_\alpha: \alpha \in A\}$ be a set of minimal normal subgroups of G and let J be the subgroup generated by the M_α. If B is a non-empty subset of A, we will say that B is *independent* if the subgroup generated by the M_β with $\beta \in B$ is just their direct product. The set $\mathscr{S}$ of all independent subsets of A is non-empty and $\mathscr{S}$ contains all unions of chains in $\mathscr{S}$. Hence $\mathscr{S}$ contains a maximal element, say B, by Zorn's Lemma. Let K be the subgroup generated by the M_β with $\beta \in B$; then of course K is actually the direct product of these M_β. Suppose that $J \neq K$; then there is an $\alpha \in A$ such that $M_\alpha \nleq K$. Since $K \lhd G$, the minimality of M_α leads us to the conclusion $K \cap M_\alpha = 1$. Hence $KM_\alpha = K \times M_\alpha$ and $B \cup \{\alpha\} \in \mathscr{S}$, which contradicts the maximality of B. Therefore $J = K$. ∎

Corollary 1 (Remak [1], p. 4). The socle of a group G is the direct product of a (possibly empty) set of minimal normal subgroups of G. The socle satisfies Min-G if and only if it is the direct product of finitely many minimal normal subgroups of G.

If M is a minimal normal subgroup of a group G, then, of course, M is characteristically simple, so either it is perfect with trivial centre or it is abelian. In the latter event M is either an elementary abelian p-group for some prime p or a direct product of copies of the additive group of rational numbers.

Suppose that M is a minimal normal subgroup of a group G and that M satisfies Min-n; then M contains a minimal normal subgroup N and $M = N^G$. Lemma 5.23 shows that M is the direct product of finitely many conjugates of N, including N itself we may suppose. Let L be a non-trivial normal subgroup of N; then $L \lhd M$ since N is a direct factor of M, and therefore $L = N$. Consequently N is simple. We state this as

Corollary 2. Let M be a minimal normal subgroup of a group. Then M satisfies Min-n if and only if M is the direct product of finitely many isomorphic simple groups.

In general, however, it is quite possible for a minimal normal subgroup to be indecomposable without being simple: see Part 2, p. 81, example (iii).

Lemma 5.23 implies that the product of all the abelian minimal normal subgroups of a group G is abelian; this subgroup is called the *abelian socle* of G, and it is easy to see that *the socle is the direct product of the abelian socle and the non-abelian socle*, i.e. the subgroup generated by all the non-abelian minimal normal subgroups of G (see Scott [5], p. 169, 7.4.13).

The Upper Socle Series

The *upper socle series* of a group G is the ascending characteristic series $\{S_\alpha\}$ defined in the following way:

$$S_0 = 1 \text{ and } S_{\alpha+1}/S_\alpha = \text{the socle of } G/S_\alpha.$$

The term "ascending Loewy series" has also been used—see Zassenhaus [3], p. 216. This series terminates after a finite or infinite number of steps but, of course, it may not reach G. However if G satisfies Min-n, each non-trivial homomorphic image of G has non-trivial socle, so G coincides with some S_α. Also the factor $S_{\beta+1}/S_\beta$ is the direct product of finitely many minimal normal subgroups of G/S_β for each β.

It is not known if the converse of this is true: *if* $G = S_\alpha$ *and if each* $S_{\beta+1}/S_\beta$ *is a direct product of finitely many minimal normal subgroups of* G/S_β, *does it follow that* G *satisfies Min-n*? This question was first raised by Baer ([13], p. 3): see also Černikov [26] or [29], § 5. We shall prove a positive result which is somewhat weaker.

Theorem 5.24 (Baer [13], p 3). A group G satisfies Min-n if and only if the socle of each non-trivial homomorphic image H of G is non-trivial and is the direct product of finitely many minimal normal subgroups of H.

Proof. Let G have the second property of the theorem and let S_α be the αth term of the upper socle series of G. Then $S_{\alpha+1}/S_\alpha$ satisfies Min-G since this property is $\boldsymbol{P}$-closed. If S_β is the last term of the upper socle series, G/S_β must have trivial socle and hence $G = S_\beta$. Let $1 \neq N \lhd G$; then there is a first ordinal α such that $N \cap S_\alpha \neq 1$ and α cannot be a limit ordinal, so $N \cap S_{\alpha-1} = 1$. Now the natural isomorphism

$$N \cap S_\alpha \simeq (N \cap S_\alpha)\, S_{\alpha-1}/S_{\alpha-1}$$

is a G-operator isomorphism and $(N \cap S_\alpha)\, S_{\alpha-1}/S_{\alpha-1}$ inherits from $S_\alpha/S_{\alpha-1}$ the property Min-G. Hence $N \cap S_\alpha$ has Min-G and consequently $N \cap S_\alpha$ contains a minimal normal subgroup of G. Thus $N \cap S_1 \neq 1$ and $\alpha = 1$.

Suppose that $N_1 > N_2 > \dots$ is an infinite, strictly descending sequence of normal subgroups of G and let I be the intersection of the N_i. Then we can suppose that $I = 1$, for G/I inherits the hypothesis from G. Now S_1 satisfies Min-G and $N_i \cap S_1 \lhd G$; hence there is an integer i such that $N_i \cap S_1 = N_{i+1} \cap S_1 =$ etc. and this implies that $N_i \cap S_1$ is trivial, for it lies inside every N_j. By the first part of the proof it follows that $N_i = 1$. This contradiction shows that G satisfies Min-n. ∎

Corollary (Černikov [26], Theorem 5.2). A group is a Černikov group if and only if the socle of each non-trivial homomorphic image is finite and non-trivial.

Proof. The condition is necessary since a minimal normal subgroup of a Černikov group is finite and since Černikov groups have Min. Conversely, let G satisfy the condition: then G satisfies Min-n by the theorem. Since G is also hyperfinite, Corollary 2 to Theorem 5.21 shows that G is a Černikov group. □

For further results about the socle of a group the reader is referred to Černikov [26], § 5, and Baer [52], § X.

Soluble Groups Satisfying Min-*n*

If G is a nilpotent group with Min-n, each upper central factor of G has Min and consequently G has Min (Jennings [1]). Thus G is a central-by-finite Černikov group by Theorem 3.14 and G' is finite by Theorem 4.12.

On the other hand it has been known for some time that there exist soluble groups which satisfy Min-n but not Min: the first example of such a group was given by Čarin [1] and subsequently similar examples were given by P. Hall (in Duguid and McLain [1]) and by Baer ([39], pp. 130—131.)

A Metabelian Group Satisfying Min-*n* but not Min

Let K be the algebraic closure of the field $GF(p)$; then K^*, the multiplicative group of K, is a direct product of quasicyclic groups, one for each prime other than p (see Fuchs [3], p. 298, Theorem 77.1). Let X be a subgroup of K^* and let F be the subfield of K generated by X. Denote by A the additive group of F and observe that multiplication by an element of X induces an automorphism in A. In fact in this way we obtain an isomorphism of X with a subgroup of Aut A. Let G be the semi-direct product of A by X: thus

$$G = AX, \quad A \lhd G, \quad A \cap X = 1.$$

Suppose that N is a non-trivial normal subgroup of G contained in A and let $0 \neq a \in N$. Then we can write

$$a^{-1} = \sum_{x \in X} \lambda_x x$$

where $\lambda_x \in GF(p)$ and all but a finite number of the λ_x are 0. Now $ax \in N$ for all $x \in X$, so N contains the element

$$\sum_{x \in X} \lambda_x (ax) = aa^{-1} = 1.$$

Hence N contains X, which implies that $N = A$. It follows that A is a minimal normal subgroup of G.

A group H is said to be *monolithic* if the intersection M of all its non-trivial normal subgroups is non-trivial: in this event M called the *monolith* of H. Evidently M is the unique minimal normal subgroup of H.

In our example G is monolithic with monolith A: for $A = C_G(A)$, so the only normal subgroup of G disjoint from A is 1.

If we choose X to be a subgroup of K^* of type q^∞, where $q \neq p$, then G satisfies Min-n since both A and G/A satisfy Min-G; but G does not satisfy Min because A is an infinite elementary abelian p-group. This group is Čarin's example. Notice that it shows that the property Min-n is not inherited by normal subgroups.

Theorem 5.25 (Baer [35], p. 390, Note II). A soluble group which satisfies Min-n is locally finite.

This is one of the few results known about soluble groups satisfying Min-n. The proof depends on the following.

Lemma 5.26 (Baer [35], pp. 385—386, Proposition). Let A be an irreducible group of automorphisms of an abelian group G. Then if A is locally finite, G is an elementary abelian p-group for some prime p.

Proof (J. E. Roseblade). Suppose that G is not elementary abelian: then, being a characteristically simple, abelian group, G is a direct product of copies of the additive group of rational numbers. Now $A \neq 1$ because $G \neq 1$, so there is a $g \in G$ for which $[g, A] \neq 1$. Since $[g, A]$ is A-admissible, $G = [g, A]$. There is an element x in G such that $g = x^2$, because G is radicable. Then $x \in [g, A]$, so $x \in [g, F]$ where F is some finitely generated, and therefore finite, subgroup of A. Thus

$$x \in [x^2, F] = [x, F]^2,$$

and consequently $[x, F] = [x, F]^2$. However, $[x, F]$ is a finitely generated, torsion-free abelian group; hence $[x, F] = 1$ and therefore $[g, F] = 1$. However this leads to the contradiction $g = 1$. ▯

Proof of Theorem 5.25. Let G be soluble and assume that it has the property Min-n. If G is abelian, it satisfies Min and the result is trivially true. Thus we can suppose that G is not abelian and use induction on the derived length of G. If N is the last non-trivial term of the derived series of G, then G/N is locally finite. Now N satisfies Min-G, so there is an ascending chief G-series in N; let F be any factor of this series. If $C = C_G(F)$, then $N \leq C$ and therefore G/C is locally finite. Also G/C is isomorphic with an irreducible group of automorphisms of the abelian group F. By Lemma 5.26 the factor F is elementary abelian and consequently N is periodic. Thus G is periodic and therefore locally finite. ▯

We remark that the paper [35] of Baer contains further results about irreducible, locally finite groups of automorphisms of abelian groups: see also Theorem 9.55 (in Part 2) for similar results.

It is apparently unknown if soluble groups which satisfy Min-n are countable, although McDougall has established the countability of metabelian groups satisfying Min-n ([2], Theorem 4.4).

Locally Soluble and Locally Nilpotent Groups Satisfying Min-*n*

Theorem 5.27 (Mal'cev [2]). The properties "every chief factor is abelian" and "every chief factor is central" are $\boldsymbol{L}$-closed.

Proof (McLain [3]). Let G satisfy the condition "every chief factor is abelian" locally and suppose that G has a non-abelian, minimal normal subgroup N; then there exist a and b in N such that $c = [a, b] \neq 1$. By minimality of N we find that $N = c^G$ and consequently that a and b belong to a subgroup $\langle c^{g_1}, \dots, c^{g_n}\rangle$ for some finite subset $\{g_1, \dots, g_n\}$. By hypothesis there exists a subgroup H containing the subset $\{c, g_1, \dots, g_n\}$ such that all chief factors of H are abelian. Now $c = [a, b] \in (c^H)'$, so $c^H = (c^H)'$. By Zorn's Lemma we can choose a subgroup M of c^H which is maximal with respect to $M \lhd H$ and $c \notin M$. Any larger H-admissible subgroup of c^H would contain c and therefore coincide with c^H. Thus c^H/M is a chief-factor of H and is abelian; hence $(c^H)' \leqq M < c^H$, which is a contradiction.

Next let G satisfy the condition "every chief factor is central" locally and suppose that G has a non-central, minimal normal subgroup N. Then there exist $a \in N$ and $g \in G$ such that $b = [a, g] \neq 1$. By minimality of N we have $N = b^G$, so $a \in \langle b^{g_1}, \dots, b^{g_n}\rangle$ for some finite subset $\{g_1, \dots, g_n\}$. There is a subgroup H all of whose chief factors are central which contains $\{g, g_1, \dots, g_n, a\}$. Set $A = a^H$; then $[A, H]$ contains b and all its conjugates in H since $[A, H]^H = [A, H]$. Therefore $a \in [A, H]$ and $A = a^H \leqq [A, H]$, which shows that $A = [A, H]$. Let M be a maximal H-admissible subgroup of A not containing a; then A/M is a chief factor of H and hence $[A, H] \leqq M < A$. ▯

Since every minimal normal subgroup of a soluble group is abelian and every minimal normal subgroup of a nilpotent group is central, we derive at once

Corollary 1 (Mal'cev [2]). A chief factor of a locally soluble group is abelian and a chief factor of a locally nilpotent group is central. Hence a simple locally soluble group is cyclic of prime order.

Corollary 2 (Čarin [4], McLain [3]). The locally nilpotent groups which satisfy Min-*n* are precisely the hypercentral Černikov-groups. Thus for locally nilpotent groups the properties Min and Min-*n* coincide.

Proof. Let G be a locally nilpotent group which satisfies Min-*n*. The property Min-*n* enables us to construct an ascending chief series in G and Corollary 1 shows that this is an ascending central series of G. That G is a Černikov group now follows by Corollary 2 to Theorem 5.21. ▯

This result was proved for locally finite p-groups by Ado [1], [3]. Notice that a locally nilpotent group satisfying Min-*n* need not be nilpotent in view of the example of the locally dihedral 2-group (p. 36). However, *a Baer group satisfying Min-n is nilpotent*: for such a group is a Černikov group and by Lemma 3.13 the finite residual lies in the centre.

The first corollary to Theorem 5.27 shows that *a locally soluble group satisfying Min-n is necessarily hyperabelian.* On the other hand, McLain [5] has constructed a periodic locally soluble group which has the property Min-n but is not soluble. It is apparently unknown if a locally soluble group which satisfies Min-n is periodic; if this were true it would provide a generalization of Baer's Theorem 5.25.

Some Further Situations when Min-n Implies Min

Theorem 5.28 (McLain [5]). Let G be a locally polycyclic group and assume that G satisfies Min-n.

(i) G is locally finite.

(ii) A necessary and sufficient condition for G to be a soluble Černikov group is that the ranks of the chief factors of finite subgroups of G be bounded.

Proof. Let N be a minimal normal subgroup of G; then N is abelian by Corollary 1 to Theorem 5.27. Suppose that N contains an element a of infinite order and set $b = a^2$. Then $N = b^G$ and hence $a \in \langle b^{g_1}, \ldots, b^{g_n}\rangle$ for some finite subset $\{g_1, \ldots, g_n\}$. Let $H = \langle b, g_1, \ldots, g_n\rangle$; then $B = b^H \leqq N$, so B is abelian. Since H is polycyclic, B is finitely generated; it is also infinite, so $B > B^2$. But $a \in B$, so $b = a^2 \in B^2$ and $B = b^H \leqq (B^2)^H = B^2$. By this contradiction N is periodic and thus every chief-factor of G is periodic. It follows that G is periodic and hence locally finite. Notice that we have shown that *a chief factor of a locally polycyclic group is elementary abelian.*

We now assume in addition that the ranks of the chief factors of finite subgroups of G are bounded by some number r. Let N be a minimal normal subgroup of G; then N is an elementary abelian p-group for some prime p. Suppose that N is infinite. Then there is set of $r + 1$ linearly independent elements of N, say $\{a_1, \ldots, a_{r+1}\}$. The subgroup $A = \langle a_1, \ldots, a_{r+1}\rangle$ is finite and also $N = a^G$ for each non-trivial element a of A. Hence there exists a finite subset $\{g_1, \ldots, g_m\}$ of G such that $A \leqq \langle a^{g_1}, \ldots, a^{g_m}\rangle$ for each non-trivial a in A. Let

$$H = \langle a_1, \ldots, a_{r+1}, g_1, \ldots, g_m\rangle$$

and notice that $A \leqq H$ and H is finite. Let $1 \neq a \in A$; then $A \leqq a^H \leqq A^H$, from which it follows that $L = a^H = A^H$ is independent of a. By Zorn's Lemma there is a subgroup M of L which is maximal with respect to the properties $M \lhd H$ and $a_1 \notin M$. Suppose that $M \cap A \neq 1$ and let

$1 \neq a \in M \cap A$. Then $L = a^H \leqq M$ and, since $A \leqq L$, this gives the contradiction $a_1 \in M$. Thus $M \cap A = 1$ and $A \simeq MA/M \leqq L/M$. But L/M is a chief factor of the finite subgroup H: for if M_1 is an H-admissible subgroup satisfying $M < M_1 \leqq L$, then $a_1 \in M_1$ and hence $L = a_1^H \leqq M_1$, i.e. $M_1 = L$. It follows that L/M, and hence A, has rank not exceeding r, which contradicts the definition of A.

Since the hypotheses on G are inherited by homomorphic images, this argument shows that each chief factor of G is finite; therefore G is hyperfinite. The property Min-n ensures that G is a Černikov group in view of Corollary 2 to Theorem 5.21, and it is clear that G is soluble. The converse is obvious. ▯

It will be seen that the proof of Theorem 5.28, in conjunction with Theorem 5.27, proves somewhat more. Let $\mathscr{P}_r$ denote the property "every chief factor is elementary abelian with rank not exceeding r." Then *a group which satisfies both $\mathscr{P}_r$ and Max locally satisfies $\mathscr{P}_r$* (McLain [5]). Taking $r = 1$ we obtain

Corollary. A locally supersoluble group satisfies Min-n if and only if it is a hypercyclic Černikov group. Thus for locally supersoluble groups the properties Min and Min-n coincide.

Theorem 5.29 (cf. Čarin [9], Theorem 7, Černikov [26], Theorem 4.1). Let G be a radical group satisfying Min-n and let N be a normal subgroup of G. Then N is a Černikov group if and only it there exists a G-admissible series in N whose factors are either FC-central with respect to G or abelian of finite rank.

From this we can derive criteria for a radical group satisfying Min-n to be a Černikov group.

Corollary 1. A radical group G satisfying Min-n is a Černikov group if and only it the Hirsch-Plotkin radical has a G-admissible series whose factors are either FC-central with respect to G or abelian of finite rank.

This follows if we take the N in Theorem 5.29 to be the Hirsch-Plotkin radical of G and then apply Theorem 3.32 (strictly speaking we need only the final argument of the proof of Theorem 3.32 here): a special case of this result is in Plotkin [16].

Since a locally soluble group with the property Min-n is hyperabelian and therefore radical, we deduce from Theorem 5.29

Corollary 2 (Čarin [4]). A locally soluble group is a Černikov group if and only if it satisfies Min-n and each chief factor has finite rank.

The proof of Theorem 5.29 requires certain information about soluble linear groups over the field of rational numbers and this we now proceed to derive.

Lemma 5.29.1 (see Robinson [7], Lemma 3.12). Let A be an irreducible abelian group of linear transformations of a finite dimensional vector space over the field of rational numbers. Then A is the direct product of a finite cyclic group and a free abelian group of countable rank.

Proof. Let A act on the n-dimensional vector space V over Q, the field of rational numbers, and let $\mathscr{R} = QA$, the group algebra of A over Q. Then V is an irreducible $\mathscr{R}$-module and $V = v\mathscr{R}$ where v is a non-zero vector in V. The mapping $\alpha \to v\alpha$ is a homomorphism of $\mathscr{R}$ onto V, regarded as right $\mathscr{R}$-modules, and the kernel $\mathscr{K}$ is a maximal ideal of the commutative ring $\mathscr{R}$; thus $\mathscr{R}/\mathscr{K} \simeq V$, as right $\mathscr{R}$-modules, and $\mathscr{R}/\mathscr{K}$ is a field. Evidently $\mathscr{R}/\mathscr{K}$ has dimension n over Q and so $\mathscr{R}/\mathscr{K}$ is an algebraic number field. The mapping $a \to a + \mathscr{K}$ is an isomorphism of A with a subgroup of the multiplicative group of $\mathscr{R}/\mathscr{K}$ and the latter group is well-known to be the direct product of a finite cyclic group and a free abelian group of countably infinite rank (Skolem [1]; see also Čarin [3], Theorem 1 and Fuchs [3], p. 297, Theorem 76.2). □

Corollary 1 (Čarin [6]). Let G be a soluble group of linear transformations of a finite dimensional vector space V over the field of rational numbers: then G has a normal unitriangulable subgroup U such that G/U is free abelian-by-finite. If G is irreducible, it is free abelian-by-finite.

Proof. Suppose first that G is irreducible. Then Mal'cev's theorem (3.21) shows that G is abelian-by-finite, and by Lemma 5.29.1 we see that G is free abelian-by-finite.* Now let G be reducible and form a QG-composition series in V (here Q is the field of rational numbers). G induces in each composition factor a free abelian-by-finite group of linear transformations. Let U be the subgroup of all elements of G that induce the identity linear transformation in each composition factor; then U is unitriangulable and G/U is free abelian-by-finite. □

Corollary 2. A soluble group of linear transformations of a finite dimensional vector space over the field of rational numbers is unitriangulable if it has no proper subgroups of finite index.

This follows at once from Corollary 1.

Corollary 3 (cf. Čarin [4], pp. 33—34). A soluble linear group over the field of rational numbers is finite if and only if it satisfies Min-n.

Proof. Let G be soluble and linear over the rational field and let G satisfy Min-n. Theorem 5.21 shows that we can assume that G has no proper subgroups of finite index and Corollary 2 implies that G is nilpotent and torsion-free. But G satisfies Min-n and hence Min, so $G = 1$. □

* Clifford's theorem is also used at this point.

Proof of Theorem 5.29. Let G be a radical group which satisfies Min-n and let $N \lhd G$; we shall assume that N has a G-admissible series whose factors are either FC-central with respect to G or abelian of finite rank and prove that N is a Černikov group.

By Min-n this is an ascending series and clearly we can suppose that each factor is either FC-central with respect to G or *torsion-free* abelian of finite rank. Let R be the finite residual of G—so R satisfies Min-n by Theorem 5.21 — and let F be a factor of the series in N. If F is FC-central in G, then R centralizes F. Suppose that F is a torsion-free abelian group of finite rank: then $R/C_R(F)$ is a radical linear group over the field of rational numbers, by the remarks preceding the proof of Theorem 3.25. Hence $R/C_R(F)$ is soluble, by the Corollary to Theorem 3.23, and Corollary 3 to Lemma 5.29.1 implies that $R/C_R(F)$ is finite. Therefore $R = C_R(F)$. It follows that the given series in N is central with respect to R and consequently that $R \cap N$ lies in the hypercentre of R. Corollary 1 to Theorem 5.21 implies that $R \cap N$ is abelian and satisfies Min. Finally, since $N/R \cap N \simeq NR/R$, which is finite, we conclude that N is a Černikov group. The converse is obvious. ▯

Further criteria for a group satisfying Min-n to be a Černikov group many be found in papers of Baer [23], [48], Robinson ([9], Theorem E*) and Newell [4].

5.3 The Maximal Condition on Normal Subgroups

The maximal condition on normal subgroups is denoted by

$$\text{Max-}n.$$

We recall that this property is $\boldsymbol{P}$, $\boldsymbol{H}$ and $\boldsymbol{R}_0$-closed (Corollary, Lemma 1.48). Max-n, like Min-n, is not inherited by subgroups or even by normal subgroups, as an example on *p*. 30 shows. However Max-n is inherited by subgroups of finite index.

Theorem 5.31 (Wilson [1]). If a group G satisfies Max-n, then every subgroup with finite index in G satisfies Max-n.

Proof. Let H be a subgroup with finite index in G: we can assume that $H \lhd G$. Suppose that H does not satisfy Max-n and let K be a normal subgroup of G which is maximal subject to H/K not satisfying Max-n.

Let $\mathscr{S}$ be the set of all non-empty finite subsets X of G with the following property: if $K = K_1 < K_2 < \cdots$ is an infinite ascending sequence of normal subgroups of H, then

$$K = \text{Core}_X K_i$$

for each integer i. One verifies easily that any transversal to H in G belongs to $\mathcal{S}$, so that $\mathcal{S}$ is not empty.

We select a minimal element X from $\mathcal{S}$ and note that there is nothing to be lost in assuming that $1 \in X$. The subset X has at least two elements, for $X = \{1\}$ would preclude the possibility of an infinite ascending normal chain in H/K. Now write

$$X^* = X \backslash \{1\}.$$

Let $K = K_1 < K_2 < \cdots$ be an infinite ascending sequence of normal subgroups of H and let

$$L_i = K_i(\mathrm{Core}_{X^*} K_i).$$

Then $K_i \leqq L_i \lhd H$ and $L_i \leqq L_{i+1}$. Suppose that $L_i = L_{i+1}$; then, since $\mathrm{Core}_X K_{i+1} = K$, we have

$$K_i = K_i(\mathrm{Core}_X K_{i+1}) = K_i(K_{i+1} \cap \mathrm{Core}_{X^*} K_{i+1}) \geqq K_{i+1} \cap L_i = K_{i+1}.$$

This is impossible, so $L_i < L_{i+1}$ for $i = 1, 2, \ldots$ and

$$\mathrm{Core}_X L_i = K$$

by the defining property of $\mathcal{S}$. Hence

$$K_i = K_i(\mathrm{Core}_X L_i) = K_i(L_i \cap \mathrm{Core}_{X^*} L_i) = L_i \cap (K_i \, \mathrm{Core}_{X^*} L_i) \geqq L_i,$$

so $K_i = L_i$ and consequently

$$\mathrm{Core}_{X^*} K_i = \mathrm{Core}_X K_i = K.$$

Hence $X^* \in \mathcal{S}$, which contradicts the minimality of X. ▯

The *FC*-Hypercentre in Groups Satisfying Max-*n*

Theorem 5.32. Let G be a group which satisfies Max-n and let $N \lhd G$. Then N is contained in the FC-hypercentre of G if and only if the following hold:

(i) N is a finite extension of a finitely generated nilpotent subgroup.

(ii) If X is a free abelian, G-admissible factor of N such that $\mathrm{Aut}_G X$ is rationally irreducible, then $\mathrm{Aut}_G X$ is finite.

Proof. Let $F_1, F_2, \ldots$ be the terms of the upper FC-central series of G. Then by Max-n it follows that F, the FC-hypercentre of G, coincides with F_m for some finite m. Also Max-n implies that $F_1 = x_1^G \ldots x_n^G$ for some finite subset $\{x_1, \ldots, x_n\}$. Now each x_i^G is finitely generated since $|G: C_G(x_i)|$ is finite; hence F_1 is finitely generated. This implies that $|G: C_G(F_1)|$ is finite and hence that F_1 is central-by-finite and satisfies Max. The same argument can be applied to each F_{i+1}/F_i. We may therefore conclude that F satisfies Max and that G/I is finite where I is the intersection of the $C_G(F_{i+1}/F_i)$, $i = 0, 1, \ldots, m-1$. Clearly $F/F \cap I$ is finite and $F \cap I$ is a finitely generated nilpotent group.

Now let N be a normal subgroup of G contained in F: then by what we have just proved, N satisfies (i). Let X be a non-trivial G-admissible factor of N which is free abelian and such that $\mathrm{Aut}_G X$ is rationally irreducible: there is no loss here in taking X to be a subgroup of N. Since $X \leq F = F_m$, there is a first integer i such that $X/X \cap F_i$ is finite; clearly $i > 0$. Thus $X/X \cap F_{i-1}$ is infinite. If $X \cap F_{i-1} \neq 1$, the rational irreducibility of $\mathrm{Aut}_G X$ implies that $X/X \cap F_{i-1}$ is periodic—and hence finite since it is finitely generated and abelian. Therefore $X \cap F_{i-1} = 1$. Consequently there is a natural G-operator isomorphism

$$X \cap F_i \simeq (X \cap F_i) F_{i-1}/F_{i-1}.$$

It follows that $X \cap F_i \leq F_1$ and $i = 1$. Thus $X/X \cap F_1$ is finite and $\mathrm{Aut}_G (X/X \cap F_1)$ is finite, as is $\mathrm{Aut}_G (X \cap F_1)$. Since X is abelian and

$$\mathrm{Hom}\,(X/X \cap F_1, X \cap F_1)$$

is finite, it follows that $\mathrm{Aut}_G X$ is finite and N satisfies (ii).

Finally let N be a normal subgroup of G which satisfies (i) and (ii). From (i) we see that N has an ascending G-admissible series whose factors are either finite or finitely generated and free abelian and such that G induces a rationally irreducible group of automorphisms in each infinite factor. But (ii) implies that G induces a finite group of automorphisms in every factor of the series, so N lies in the FC-hypercentre of G. (This is true even if G does not satisfy Max-n). ▯

Corollary (McLain [4]). The FC-hypercentral groups which satisfy Max-n are precisely the finite extensions of finitely generated nilpotent groups. Hence Max-n and Max coincide for FC-hypercentral groups.

In particular the theorem shows that the hypercentre of a group with Max-n is a finitely generated nilpotent group; thus Max-n and Max coincide for nilpotent groups (Jennings [1]), but of course this is easy to prove directly.

Soluble Groups Satisfying Max-n

The following fact is analogous to the periodicity of soluble groups with the property Min-n.

Theorem 5.33 (P. Hall [4], p. 420). A soluble group which satisfies Max-n is finitely generated.

Proof. Let G be soluble and suppose that G satisfies Max-n. If G is abelian, the assertion is obvious, so let G be non-abelian and denote by

A the least non-trivial term of the derived series of G. By induction on the derived length we may assume that G/A is finitely generated, say by $Ax_1, \ldots, Ax_m$.

Let $X = \langle x_1, \ldots, x_m \rangle$, so that $G = AX$. Since A satisfies Max-G, we can write $A = y_1^G \cdots y_n^G$ for some finite subset $\{y_1, \ldots, y_n\}$. But A is abelian, so $y_i^G = y_i^X$ and $G = AX = \langle y_1, \ldots, y_n, x_1, \ldots, x_m \rangle$. ☐

Thus for soluble groups Max-n is intermediate between the properties "Max" and "finitely generated", and we will very shortly give examples to show that it is distinct from both these properties.

On the other hand, P. Hall has proved the following result.

Theorem 5.34 (P. Hall [4], Theorem 3). A finitely generated abelian-by-polycyclic group satisfies Max-n.

Thus, in particular, *finitely generated metabelian groups satisfy Max-n.* It follows that the group

$$\langle a, b \colon b^{-1}ab = a^2 \rangle$$

satisfies Max-n: on the other hand it does not satisfy Max since the subgroup $\langle a, a^{b^{-1}}, a^{b^{-2}}, \ldots \rangle$ is not finitely generated. This example shows that the property Max-n is not $\boldsymbol{S}_n$-closed.

The proof of Theorem 5.34 depends on the following lemma on modules which the reader will recognize as a variant of Hilbert's Basis Theorem.

Lemma 5.35 (P. Hall [4], Lemma 3). Let N be a normal subgroup of a group G such that G/N is either infinite cyclic or finite and let R be a ring with identity. Let A be a (right) RG-module and let B be a (right) RN-module contained in A such that A is generated as an RG-module by B. Then if B satisfies the maximal condition on RN-submodules (Max-RN), A satisfies the maximal condition on RG-submodules (Max-RG).

Proof. Case (i) *G/N is finite.* Let $\{t_1, \ldots, t_n\}$ be a transversal to N in G. Since A is generated as an RG-module by B, we have $A = Bt_1 + \cdots + Bt_n$. If $b \in B$ and $x \in N$, then

$$bt_i x = bx^{t_i^{-1}} t_i \in Bt_i$$

since $N \lhd G$. Hence Bt_i is an RN-module and the mapping $b \to bt_i$ is a group isomorphism of B with Bt_i which maps RN-submodules of B onto RN-submodules of Bt_i. Thus Bt_i satisfies Max-RN and consequently so does A, by the $\boldsymbol{P}$ and $\boldsymbol{H}$-closure of the property Max-RN. Thus *a fortiori* A satisfies Max-RG.

Case (ii) *G/N is infinite cyclic.* Let $G = \langle x, N \rangle$; then, if $0 \neq a \in A$, we can write (not necessarily uniquely)

$$a = \sum_{i=r}^{s} b_i x^i$$

where $b_i \in B$, $r \leqq s$ and $b_s \neq 0$. The element b_s will be called a *leading coefficient* of a and $b_s x^s$ will be called a *leading term* of a.

Let A_0 be an RG-submodule of A: it will be enough to prove that A_0 is generated as an RG-module by some finite subset and obviously we may assume that $A_0 \neq 0$. Let B_0 denote the set consisting of 0 and all leading coefficients of elements of A_0. Then B_0 is an RN-submodule of B: for if a and a' belong to A_0 and have leading terms $b_s x^s$ and $b'_{s'} x^{s'}$ where $s' \leqq s$, then $a \pm a' x^{s-s'}$ has $(b_s \pm b'_{s'})\, x^s$ as a leading term, unless of course $b_s \pm b'_{s'} = 0$; also, if $u \in RN$, then $a(x^{-s} u x^s)$ belongs to A_0 and its leading term is $(b_s u)\, x^s$ unless $b_s u = 0$.

By hypothesis B_0 can be finitely generated as an RN-module, say by the non-zero elements $b_1, \ldots, b_l$. Let a_i be an element of A_0 which has b_i as a leading coefficient; clearly we can choose the a_i so that none of them involves a negative power of x and each of their leading terms involves the same positive power of x, say x^m. Thus a leading term of a_i is $b_i x^m$. Let A_1 be the RG-submodule of A generated by $a_1, \ldots, a_l$; then certainly $A_1 \leqq A_0$. Let $0 \neq a \in A_0$: for some large positive integer n the element ax^n involves no non-negative powers of x, and of course it belongs to A_0. Let cx^p be a leading term of ax^n; then $c \in B_0$, so

$$c = \sum_{i=1}^{l} b_i r_i$$

where $r_i \in RN$. If $p \geqq m$, the element

$$a' = \sum_{i=1}^{l} a_i x^{-m} r_i x^p$$

belongs to A_1, involves no non-negative powers of x and has leading term

$$\left(\sum_{i=1}^{l} b_i r_i\right) x^p = cx^p,$$

i.e. a' and ax^n have the same element as a leading term. Hence $ax^n - a'$ belongs to A_0 and involves no negative powers of x and no powers of x higher than the $(p-1)$th. If $ax^n - a'$ still involves powers of x beyond the $(m-1)$th, the same argument may be employed with $ax^n - a'$ in place of ax^n. Thus after at most $p - m + 1$ applications of the argument we find that $ax^n \equiv a^* \bmod A_1$ where

$$a^* \in B^* = B + Bx + \cdots + Bx^{m-1}.$$

Hence $ax^n \in (B^* \cap A_0) + A_1$. Now B^* clearly satisfies Max-RN, so $B^* \cap A_0$ can be finitely generated as an RN-module, say by $a_{l+1}, \ldots, a_k$. It follows that a belongs to the RG-submodule generated by

$a_1, \ldots, a_l, \ldots, a_k$, and since these elements are independent of a, they generate A_0 as an RG-module. □

Corollary. Let R be a ring with identity which satisfies the maximal condition on right ideals and let G be a polycyclic-by-finite group. Then RG, the group ring of G over R, satisfies the maximal condition on right ideals.

We remark that it is easy to show that *if G is a group whose integral group ring satisfies the maximal condition on right ideals, then G satisfies Max* (P. Hall [4], p. 421). If every group which satisfies Max were polycyclic-by-finite, the converse of this statement would be true in view of the previous corollary.

Proof of Theorem 5.34. Let G be a finitely generated abelian-by-polycyclic group; then there is a normal abelian subgroup A of G such that G/A is polycyclic. By Lemma 1.43 and its corollary, G/A is finitely presented and $A = a_1^G \cdots a_n^G$ for some finite subset $\{a_1, \ldots, a_n\}$. By the $\boldsymbol{P}$-closure of the property Max-G, it is enough to prove that a^G satisfies Max-G for each $a \in A$.

Let $C = C_G(a^G)$; then $A \leq C$ and $H = G/C$ is polycyclic. By the Corollary to Lemma 5.35 the integral group ring of H, say R, satisfies the maximal condition on right ideals. Let $r \in R$ and write

$$r = n_1h_1 + \cdots + n_th_t$$

where $h_i = Cg_i \in H$ and the n_i are integers. If $b \in a^G$, we define

$$b^r = \prod_{i=1}^{t} (b^{n_i})^{h_i} = \prod_{i=1}^{t} (b^{n_i})^{g_i},$$

and observe that this is well-defined; then, apart from the multiplicative notation, a^G is an R-module. The mapping $r \to a^r$ is a homomorphism of right R-modules from R onto a^G. Hence a^G satisfies Max-R, the maximal condition on right R-submodules. But the R-submodules of a^G are precisely the G-admissible subgroups of a^G. Therefore a^G satisfies Max-G as required. □

Corollary (P. Hall [4], p. 430, Corollary 2). There exist only countably many non-isomorphic finitely generated groups in the class $\mathfrak{A}(\boldsymbol{P}\mathfrak{C})\,\mathfrak{F}$.

Proof. Let G be a finitely generated group in the class $\mathfrak{A}(\boldsymbol{P}\mathfrak{C})\,\mathfrak{F}$ and let G be the image of a finitely generated free group F under a homomorphism with kernel R. It is easy to see that G has a normal abelian subgroup A such that G/A is polycyclic-by-finite. If B is the inverse image of A, then $F/B \simeq G/A$ and $B/R \simeq A$. Thus F/B is finitely presented and $B = x_1^F \cdots x_k^F$ for some finite subset $\{x_1, \ldots, x_k\}$; hence there are only countably many choices for B. Next, $B' \leqq R$ and F/B' satisfies Max-n, by Theorem 5.34, so $R = y_1^F \cdots y_l^F B'$ for some finite subset $\{y_1, \dots, y_l\}$. Hence there are only countably many choices for R and so for G. $\square$

In particular there are only countably many non-isomorphic finitely generated abelian-by-nilpotent groups and polycyclic-by-finite groups.

Remark. We mention in passing that *there are only countably many non-isomorphic Černikov groups* (Wehrfritz [8]). Here is a sketch of the proof of this result.

Let G be a Černikov group and let A be its finite residual. The possibilities for A and $H = G/A$, up to isomorphism, are countable in number, so it will be sufficient if we can show that the number of equivalence classes of extensions of A by H is countable. To do this we need only show that A can be made into an H-module in countably many ways; for a factor set of A by H is a subset of A with at most $|H|^2$ elements and so the possibilities for it are countable in number. We can assume that A is the direct product of n groups of type p^∞. The proof now reduces to showing that H has countably many non-equivalent representations over R_p, the ring of p-adic integers. Let F_p be the field of p-adic numbers. By a theorem of integral representation theory an F_p-equivalence class of R_p-representations of H decomposes into finitely many R_p-equivalence classes (Curtis and Reiner [1], p. 533, Corollary 76.10). The result now follows from the fact that a finite group has only finitely many inequivalent representations of given degree over a field of characteristic 0.

The following theorem stands in contrast to the Corollary to Theorem 5.34.

Theorem 5.36 (P. Hall [4], Theorem 6). Let A be a non-trivial countable abelian group. Then there exist uncountably many non-isomorphic 2-generator groups H such that $[H'', H] = 1$, $H/\zeta(H)$ has trivial centre and

$$\zeta(H) \simeq A.$$

Proof. Let X be the group generated by the set of elements

$$\{x_i : i = 0, \pm 1, \pm 2, \dots\}$$

subject to the defining relations

$$[x_i, x_j, x_k] = 1 \tag{4}$$

and

$$[x_i, x_j] = [x_{i+k}, x_{j+k}] \tag{5}$$

where $i, j, k = 0, \pm 1, \pm 2, \dots$ Then X' is free abelian with the set of all

$$d_r = [x_i, x_{i+r}], \quad r = 1, 2, \dots,$$

as basis and X is a torsion-free nilpotent group of class 2. Notice that d_r is independent of i in view of equation (5).

The mapping $x_i \to x_{i+1}$ extends to an automorphism t of X since it preserves the relations (4) and (5). Let G be the holomorph of X by the infinite cyclic group $\langle t \rangle$. Then

$$G = \langle t, x_0 \rangle.$$

It follows from (5) that $d_r^t = d_r$ for all $r > 0$, so $X' \leqq \zeta(G)$. Since G/X' is metabelian, $G'' \leqq X'$ and consequently

$$[G'', G] = 1.$$

The group G/X' is isomorphic with the standard wreath product of two infinite cyclic groups. Hence G/X' has trivial centre and consequently

$$X' = \zeta(G).$$

X' is a free abelian group of countably infinite rank and A is a non-trivial countable abelian group, so $A \simeq X'/M$ for some M. Since $X' = \zeta(G)$, it follows that $M \lhd G$. Now let

$$H = G/M.$$

Then $\zeta(G) = X'/M \simeq A$, because G/X' has trivial centre.

It remains to show how we can obtain uncountably many non-isomorphic groups H by varying M within X'. We note first of all that there are uncountably many subgroups M such that $X'/M \simeq A$, for there are uncountably many ways of mapping X' onto A. On the other hand, there are only countably many homomorphisms of G onto H, for G is finitely generated and H is countable. Hence an uncountably infinite number of the M must give rise to non-isomorphic groups H. $\square$

If A is a non-finitely generated, countable abelian group, we see that the group H cannot satisfy Max-n. Hence we derive the following

Corollary. There exists a 2-generator group H such that $[H'', H] = 1$ but which does not satisfy Max-n. The centre of H can be any prescribed countable, but non-finitely generated, abelian group.

We remark that the group G/X'—which is isomorphic with the wreath product of two infinite cyclic groups—is not finitely presented: for if it were, X' would be finitely generated, by Lemma 1.43. Consequently, *there exist finitely generated metabelian groups which are not finitely presented.* In this connection we mention that Šmel'kin [5] has proved that *no free soluble group of rank* >1 *can be finitely presented.*

However it is still unknown if a finitely presented soluble group need satisfy Max-n: this is equivalent to asking whether every homomorphic image of a finitely presented soluble group is finitely presented.

The paper [1] of Lennox and Roseblade adds considerably to our knowledge of finitely generated abelian-by-nilpotent groups. For example, these authors prove that *subgroups of a finitely generated abelian-by-nilpotent group have boundedly finite hypercentral length and the same holds also for subgroups of groups in the class* $\mathfrak{N}^2\mathfrak{F}$ *which satisfy Max-n* (pp. 399–401): they also prove that *finitely generated abelian-by-nilpotent groups satisfy the maximal condition on centralizers of subgroups* (Theorem C).

Further papers which may be consulted are C. K. Gupta [3] and Dark and Rhemtulla [1].

Locally Soluble and Locally Nilpotent Groups Satisfying Max-*n*

Theorem 5.37 (McLain [3]). A locally nilpotent group satisfies Max-*n* if and only if it is a finitely generated nilpotent group. Thus Max and Max-*n* coincide for locally nilpotent groups.

Proof. Let G be locally nilpotent and suppose that G satisfies Max-*n*. Clearly G/G' is finitely generated, so $G = XG'$ where X is a finitely generated—and therefore nilpotent—subgroup. Let c be the nilpotent class of X; then $G = X\gamma_{c+2}(G)$ by Lemma 2.22. Hence

$$G/\gamma_{c+2}(G) \simeq X/X \cap \gamma_{c+2}(G),$$

so that $G/\gamma_{c+2}(G)$ is nilpotent of class $\leq c$ and consequently

$$\gamma_{c+1}(G) = \gamma_{c+2}(G) = L,$$

say; clearly $L = [L, G]$. If $L = 1$, then $G = X$ and G is finitely generated and nilpotent, so let us assume that $L \neq 1$. Then by Max-*n* there exists a normal subgroup M of G which is maximal with respect to being properly contained in L. Thus L/M is a chief factor of G. But Corollary 1 to Theorem 5.27 asserts that L/M is a central factor of G, and this yields the contradiction $L = [L, G] \leq M < L$. ▯

It is well-known that a maximal subgroup of a nilpotent group is necessarily normal and hence of prime index; this follows, for example, from the subnormality of an arbitrary subgroup of a nilpotent group. We shall generalize this theorem to locally nilpotent groups.

Theorem 5.38 (Baer [7], Theorem 4.1, McLain [3]). A maximal subgroup of a locally nilpotent group is normal.

Proof (McLain [3]). Suppose that M is a non-normal maximal subgroup of a locally nilpotent group G. Then certainly $G' \nleq M$; let $x \in G' \backslash M$ and write

$$x = \prod_{i=1}^{m} [y_i, z_i].$$

Since $x \notin M$ and M is maximal, $G = \langle x, M \rangle$ and consequently the y_i and z_i belong to a subgroup $H = \langle x, a_1, \ldots, a_n \rangle$ where $a_j \in M$. Now $x \notin \langle a_1, \ldots, a_n \rangle$, so by Zorn's Lemma there is a subgroup K of H which is maximal subject to $\langle a_1, \ldots, a_n \rangle \leq K$ and $x \notin K$. A subgroup of H which contains K properly also contains x and therefore equals H, i.e. K is a maximal subgroup of H. Now H is finitely generated and nilpotent, so $K \lhd H$ and therefore $H' \leq K$. However $x \in H'$, so $x \in K$, which is contrary to our choice of K. ▯

It will be seen that the proof actually shows somewhat more. *Let $\mathfrak{X}$ denote the class of all groups G such that a maximal subgroup of a subgroup of G is always normal in that subgroup; then $\mathfrak{X}$ is* **L**-*closed* (Baer [7], Theorem 4.1). In Section 6.1 (Part 2) such groups are studied under the name *$\tilde{N}$-groups*.

On the other hand, a maximal subgroup of a soluble group need not be normal or even of finite index: in Čarin's example of a metabelian group satisfying Min-n (p. 153) X is a maximal subgroup of infinite index. Some classes of groups in which maximal subgroups have finite index are discussed in Sections 9.3 and 9.5 and also in Phillips, Robinson and Roseblade [1], where a generalization of Theorem 5.38 may be found.

A Locally Soluble Group Satisfying Max-*n* which is Insoluble

We will now describe a construction due to McLain [5] of a periodic locally soluble group which satisfies Max-n yet which is insoluble; this group cannot therefore be finitely generated, which is in contrast to Theorem 5.33.

For each positive integer i we construct a finite soluble group G_i with a unique minimal normal subgroup N_i. To start the construction we take G_1 to be the symmetric group of degree 3 and N_1 its unique Sylow 3-subgroup. Suppose that G_i has been constructed and select any prime p not dividing the order of G_1; let R be the group algebra of G_i over $GF(p)$. Then R, considered as a right R-module, is completely reducible by Maschke's Theorem. If each irreducible submodule of R is non-faithful, then N_i lies in the kernel of each of the corresponding irreducible representations of G_i; therefore R is not faithful. This is wrong, so there must exist a faithful irreducible submodule of R, say N_{i+1}. Regard G_i as a group of automorphisms of the additive group N_{i+1} and let G_{i+1} be the semi-direct product of N_{i+1} with G_i. Observe that G_{i+1} is a finite soluble group and that N_{i+1} is its own centralizer in G_{i+1}. Hence N_{i+1} is the unique minimal normal subgroup of G_{i+1}.

Let G be the union of the chain $G_1 < G_2 < \cdots$. Then G is a periodic locally soluble group. Let N be a proper non-trivial normal subgroup of G. If $N \cap G_i \neq 1$, then $N_i \leq N$ since N_i is the unique minimal normal subgroup of G_i, and if $i > 1$,

$$N \cap G_i = N \cap (N_i G_{i-1}) = N_i(N \cap G_{i-1}). \qquad (6)$$

Let j be the least integer for which $N \cap G_j \neq 1$. Then N is generated by $N \cap G_j$, $N \cap G_{j+1}, \ldots$, and hence by $N \cap G_j$, N_{j+1}, N_{j+2}, ..., in view of (6). If $j > 1$, then $N \cap G_j = N_j$ by (6) and if $j = 1$, either $N \cap G_1 = N_1$ or else $N \geqq G_1$, which clearly implies that $N = G$. It follows that

$$N = \langle N_j, N_{j+1}, \ldots \rangle,$$

which is just $G^{(j)}$, the $(j+1)$th term of the derived series of G. The only non-trivial normal subgroups of G are therefore the terms of the derived series of G and G satisfies Max-n. However G is not soluble because $G^{(j)}$ is never trivial if j is finite.

We can still ask for additional conditions which will ensure that a locally soluble group satisfying Max-n be soluble. Every soluble group G has a normal series of finite length with nilpotent factors, for example the derived series. The length of a shortest series of this type is called the *nilpotent length of* G. When G satisfies Max-n, the nilpotent length of G is just the length of the upper nilpotent (or Fitting) series of G.

Groups whose finitely generated subgroups are soluble with bounded nilpotent length form a special class of locally soluble groups, as the following result of McLain demonstrates; see Rhemtulla [3] for another theorem on this class of locally soluble groups.

Theorem 5.39 (McLain [5]) Let G be a locally soluble group satisfying Max-n. Then G is soluble (and hence finitely generated) if and only if finitely generated subgroups of G have bounded nilpotent length.

To prove this we need a simple lemma.

Lemma 5.39.1. Let G be a locally soluble group which satisfies Max-n. To each non-negative integer p there corresponds a non-negative integer $m = m(p, G)$ such that

$$G^{(m)} \leqq \gamma_{r_p}(\cdots \gamma_{r_2}(\gamma_{r_1}(G)) \cdots)$$

for all sequences of positive integers $r_1, r_2, \ldots, r_p$.

Proof. Define $m(0, G) = 0$; in this case the result is vacuously true. Assume that the result has been proved for p and set $m = m(p, G)$. $G/G^{(m+1)}$ is a soluble group satisfying Max-n, so it is finitely generated, by Theorem 5.33. Hence $G = XG^{(m+1)}$ where X is a finitely generated, and therefore soluble, subgroup. Now let $r_1, \ldots, r_{p+1}$ be a sequence of positive integers. By Lemma 2.22 the equation

$$G = XG^{(m+1)} = X(G^{(m)})'$$

implies that

$$G = X\gamma r_{p+1}(G^{(m)}). \tag{7}$$

Let d be the derived length of X; then by (7) and induction on p

$$G^{(d)} \leqq X^{(d)}\gamma_{r_{p+1}}(G^{(m)}) \leqq \gamma_{r_{p+1}}(\gamma_{r_p}(\cdots(\gamma_{r_1}(G))\cdots)).$$

Finally, define $m(p+1, G) = d$. ▯

Proof of Theorem 5.39. Let G be a locally soluble group, let G satisfy Max-n and assume that all finitely generated subgroups of G have nilpotent length $\leqq l$. By Lemma 5.39.1 there is an integer m such that

$$G^{(m)} \leqq \gamma_{r_l}(\cdots\gamma_{r_1}(G)\cdots)$$

for all sequences of positive integers $r_1, \ldots, r_l$. Now $G/G^{(m+1)}$ is finitely generated, so $G = XG^{(m+1)}$ where X is finitely generated and soluble. Since X has nilpotent length at most l, there exist positive integers $r_1, \ldots, r_l$ such that

$$\gamma_{r_l}(\cdots\gamma_{r_1}(X)\cdots) = 1.$$

Hence

$$G^{(m)} \leqq \gamma_{r_l}(\cdots\gamma_{r_1}(G)\cdots) \leqq \gamma_{r_l}(\cdots\gamma_{r_1}(X)\cdots)\, G^{(m+1)} = G^{(m+1)}.$$

Thus $L = G^{(m)} = G^{(m+1)}$ and $L = L'$. If $L = 1$, then G is soluble. If $L \neq 1$, then by Max-n there exists a normal subgroup M of G which is maximal with respect to being properly contained in L. Then L/M is a chief factor of G and hence is abelian, by Corollary 1 to Theorem 5.27. But this implies that $L' \leqq M < L$, which contradicts $L = L'$. Hence G is soluble. The converse is obvious. ▯

Corollary. A locally supersoluble group satisfies Max-n if and only if it is supersoluble. Hence Max and Max-n coincide for locally supersoluble groups.

Proof. Since the automorphism group of a cyclic group is abelian, the derived subgroup of a supersoluble group centralizes every normal cyclic factor and is therefore nilpotent. Thus every supersoluble group is nilpotent-by-abelian and has nilpotent length $\leqq 2$. The theorem can now be applied directly. ▯

Primary Decompositions in Groups Satisfying Max-*n*

In [1] Kurata, generalizing earlier work of Schenkman [5], has developed a theory of primary decompositions of normal subgroups of groups with the property Max-n which is analogous to the well-known theory of primary decompositions of ideals in a commutative Noetherian ring (see, for example, Northcott [1], Chapter 1). The starting point for both Schenkman's and Kurata's investigations is the observation that the product and commutator subgroup of two normal subgroups in a group behave analogously to the sum and product respectively of two ideals in a commutative ring.

Let G be a group which satisfies Max-n: a normal subgroup N of G is said to be *prime* if $[a^G, b^G] \leq N$ implies that a or b belongs to N. This is equivalent to N not being the intersection of two normal subgroups of G properly containing it and G/N having no non-trivial normal abelian subgroups. A normal subgroup N is said to be *primary* if $[a^G, b^G] \leq N$ and $a \notin N$ imply that b belong to S, the intersection of all the minimal prime subgroups containing N: Kurata shows that S/N is just the soluble radical of G/N (Kurata [1], Proposition 4.5).

The main result of Kurata's theory is the following. *Let G be a group satisfying Max-n; then every normal subgroup of G can be represented as an irredundant intersection of finitely many primary subgroups if and only if G has the following property: given $X \lhd G$ and $Y \lhd G$, there is an integer n such that $X^{(n)} \cap Y \leq [X, Y]$*: moreover there is a uniqueness theorem for such decompositions (Kurata [1], Theorem 5.4 and Theorem 6.1).

5.4 Some Finiteness Conditions on Subnormal Subgroups

We begin by surveying the more important finiteness conditions which can be imposed on the set of subnormal subgroups of a group; then we will consider the minimal condition on subnormal subgroups in greater detail.

(i) The Maximal and Minimal Conditions on Subnormal Subgroups

We denote these properties by

Max-sn and Min-sn

respectively and observe that they are closed with respect to the operations

$$\boldsymbol{S}_n, \boldsymbol{H}, \boldsymbol{P}, \boldsymbol{R}_0,$$

either for obvious reasons or by Lemma 1.48. The $\boldsymbol{S}_n$ and $\boldsymbol{P}$-closure show that for soluble (or even FC-soluble) groups Max and Max-sn coincide, as do Min and Min-sn. In view of the examples of soluble groups with Max-n but not Max and Min-n but not Min and the existence of simple groups without Max or Min (see Theorem 5.11), we can assert that Max-sn comes strictly between the properties Max and Max-n and Min-sn comes strictly between Min and Min-n.

It is well-known that the groups which satisfy both Max-*sn* and Min-*sn* are just the groups which posses a subnormal composition series of finite length (see Kuroš [9)], § 16).

The locally soluble groups which satisfy Min-sn and the radical groups which satisfy Min-sn are just the soluble Černikov groups: for such groups are hyperabelian, by Corollary 1, Theorem 5.27, and Theorem 3.12 may be applied. *A radical group satisfies Max-sn if and only if it is polycyclic*, by Theorem 5.37; but it seems to be unknown if a locally soluble group with the property Max-*sn* need be polycyclic.

(ii) Groups in which every Subnormal Subgroup is Finitely Generated

These are the groups which constitute the class

$$\mathfrak{G}^{\boldsymbol{S}_n};$$

this is evidently an $\boldsymbol{S}_n$, $\boldsymbol{H}$, $\boldsymbol{P}$ and $\boldsymbol{R}_0$-closed class of groups. The relation of the class $\mathfrak{G}^{\boldsymbol{S}_n}$ to Max-*sn* is obscure at present; for it is not known whether a group all of whose subnormal subgroups are finitely generated need satisfy Max-*sn*. It is not difficult to see that this problem is equivalent to asking if the union of any chain of subnormal subgroups in a $\mathfrak{G}^{\boldsymbol{S}_n}$-group is subnormal.

It is known that $\mathfrak{G}^{\boldsymbol{S}_n}$ is a subnormal coalition class: more generally, if $\mathfrak{X} = \boldsymbol{S}_n\mathfrak{X} = \boldsymbol{N}_0\mathfrak{X}$, then $\mathfrak{G} \cap \mathfrak{X}$ is a subnormal coalition class (Roseblade and Stonehewer [1], Theorem A). For related results about the class $\mathfrak{G}^{\boldsymbol{S}_n}$ see Robinson [2], Theorem 5.2*.

We recall that a group is *just-infinite* if it is infinite yet all its proper homomorphic images are finite. The class of finitely generated just-infinite groups is of some interest in itself in view of the following fact: *a finitely generated infinite group G has a just-infinite homomorphic image* (Baer [16], § 1, Lemma 1). A generalization of this result is proved in Part 2, Chapter 6 (Lemma 6.17).

The connection with groups in which every subnormal subgroup is finitely generated is the following.

Theorem 5.41. In a finitely generated just-infinite group every subnormal subgroup is finitely generated.

Proof. Assume that the theorem is false. Let G be a finitely generated just-infinite group and let r be the least positive integer such that G contains a non-finitely generated subgroup which is subnormal in G in r steps; for brevity let us call this a $\lhd^r$-*subgroup*.

Suppose first of all that $r \leqq 2$; then there is normal subgroup K of G which has a non-finitely generated normal subgroup. Since the union of a chain of non-finitely generated subgroups cannot be finitely generated, Zorn's Lemma implies that there exists a normal subgroup H of K which is maximal with respect to not being finitely generated. Obviously $K \neq 1$, so G/K is finite and H has just a finite number (say m) of conjugates in G. Let

$$C = \mathrm{Core}_G H.$$

Then certainly $C \lhd G$, so either $C = 1$ or G/C is finite and in either case C is finitely generated. Let i be a positive integer less than m and assume that all intersections of $i + 1$ distinct conjugates of H are finitely generated. Let I be the intersection of i distinct conjugates of H; we will prove that I is finitely generated. Clearly we may suppose that $I \neq C$, so that there is a conjugate H_1 of H such that $I \nleqq H_1$. Then $H_1 < H_1 I \lhd K$; hence $H_1 I$ is finitely generated, by maximality of H. Also $H_1 \cap I$ is the intersection of $i + 1$ distinct conjugates of H and so is finitely generated. It now follows that I is finitely generated. Therefore, by induction on $m - i$, we find that every such I is finitely generated. Hence H is finitely generated.

By this contradiction $r > 2$ and consequently $X \lhd^2 G$ implies that X is finitely generated; therefore each normal subgroup of G satisfies Max-n.

There is a normal subgroup K of G which has a normal subgroup H possessing a non-finitely generated $\lhd^{r-2}$-subgroup. Since K satisfies Max-n, we can choose H to be maximal subject to having a non-finitely generated $\lhd^{r-2}$-subgroup. Once again H has just a finite number of conjugates in G (say m). Set

$$C = \mathrm{Core}_G H$$

and suppose that $C \neq 1$, so that G/C is finite. Let S be a non-finitely generated $\lhd^{r-2}$-subgroup of H; then $S \cap C$ is not finitely generated because $|S : S \cap C|$ is finite. But $S \cap C \lhd^{r-2} H \cap C = C$ and $C \lhd G$, so $S \cap C \lhd^{r-1} G$, which contradicts the minimality of r. Hence $C = 1$.

Let i be a positive integer less than m and assume that each $\lhd^{r-2}$-subgroup of an intersection of $i + 1$ distinct conjugates of H is finitely generated. Denote by I the intersection of a set of i distinct conjugates of H. Let $S \lhd^{r-2} I$; we will show that S is finitely generated. Assuming that $I \neq 1$, we can find a conjugate H_1 of H such that $I \nleqq H_1$; then $H_1 < H_1 I \lhd K$. Now the relations $S \lhd^{r-2} I \leqq K$ and $H_1 \lhd K$ imply that $H_1 S \lhd^{r-2} H_1 I$; thus by maximality of H the group $H_1 S$ is finitely generated. Also $H_1 \cap S \lhd^{r-2} H_1 \cap I$; hence $H_1 \cap S$ is finitely generated and therefore S is finitely generated. By induction on $m - i$ every intersection of i distinct conjugates of H has its $\lhd^{r-2}$-subgroups finitely generated; taking $i = 1$ we obtain a contradiction. ▯

Corollary. An infinite group each of whose infinite homomorphic images contains a non-finitely generated subnormal subgroup cannot be finitely generated.

An example of a finitely generated just-infinite group is the central factor group of the special linear group $SL(n, Z)$ where $n \geqq 3$ and Z is the ring of integers. For it follows from results of Brenner [2] and Mennicke [1] that a non-central normal subgroup of $SL(n, Z)$ contains a congruence subgroup and therefore has finite index: see also Bass, Lazard and Serre [1]. Now $SL(n, Z)$ is finitely generated and its centre is finite, so *subnormal subgroups of* $SL(n, Z)$ *are finitely generated,* by Theorem 5.41. Recently J. S. Wilson has extended this by showing that *non-central subnormal subgroups of* $SL(n, Z)$ *have finite index.* Many new results on just-infinite groups are contained in the paper [3] of Wilson.

(iii) Groups with Bounded Subnormal Indices

If H is a subnormal subgroup of a group G, the *subnormal index* or *defect* of H in G

$$s(G: H)$$

is the least non-negative integer s such that $H \lhd^s G$; thus s is the length of a shortest series joining H to G. Clearly $s(G: H) = 0$ and 1 are equivalent to $H = G$ and to H being a proper normal subgroup of G respectively.

If H is an arbitrary subset of G, *the series of successive normal closures* of H in G is the descending series $\{H^{G,\alpha}\}$ between H and G defined by

$$H^{G,0} = G, \quad H^{G,\alpha+1} = H^{H^{G,\alpha}} \quad \text{and} \quad H^{G,\lambda} = \bigcap_{\beta<\lambda} H^{G,\beta}$$

where α is an ordinal and λ a limit ordinal. It is well-known that this is the fastest descending series whose terms all contain H, and that $H \lhd^s G$ is equivalent to $H = H^{G,s}$ (see Robinson [11], § 3, for example).

It is easy to show by induction on i and use of Lemma 2.11 that

$$H^{G,i} = H[G, \underbrace{H, \ldots, H}_{\longleftarrow i \longrightarrow}]$$

for each non-negative integer i.

If i is a non-negative integer, we define

$$\mathfrak{B}_i$$

to be the class of groups in which no subnormal index exceeds i and

$$\mathfrak{B} = \bigcup_{i=0,1,2,\ldots} \mathfrak{B}_i,$$

the class of groups with bounded subnormal indices.

Suppose H is a subnormal subgroup of a group G and

$$s(G: H) = s + 1 > 0.$$

Then

$$H = H^{G,s+1} < H^{G,s} < \cdots < H^{G,0} = G,$$

and, by definition of the subnormal index, $s(G: H^{G,s}) = s$. It follows that G has a subnormal subgroup with subnormal index s if it has one with subnormal index $s + 1$. Hence *either* $G \in \mathfrak{B}_i$ *for some* i *and* G *has subnormal subgroups with subnormal indices* $0, 1, \ldots, i$ *or* G *has subnormal subgroups of every non-negative subnormal index.*

$\mathfrak{B}_1$ is just the class of groups in which normality is a transitive relation—or $\mathfrak{T}$*-groups* as these are sometimes called. $\mathfrak{T}$-groups, particularly when soluble or locally finite, have been widely studied in recent years. The relevant papers are Abramovskiĭ [1], [2], Abramovskiĭ and Kargapolov [1], Best and Taussky [1], Gaschütz [1], Peng [2], Robinson [1], [8], [13], Zacher [1]; for more general classes see Medvedeva [1], Menogazzo [1], [2] and Weidig [1]. As a sample of the results obtained we mention the following: *soluble* $\mathfrak{T}$*-groups are metabelian* and *finitely generated soluble* $\mathfrak{T}$*-groups are finite or abelian* (Robinson [1], Theorems 2.3.1 and 3.3.1): a *locally finite group in which every 2-generator soluble subgroup with derived length* $\leqq 3$ *and order divisible by at most two primes is a* $\mathfrak{T}$*-group is itself a* $\mathfrak{T}$*-group* (Robinson [13]). Finite soluble $\mathfrak{T}$-groups were first classified by Gaschütz [1] and Gaschütz's classification was later extended by Robinson to include all countable periodic soluble $\mathfrak{T}$-groups ([1], Theorems 5.1.2 and 5.2.1). An interesting unsolved problem is whether a locally nilpotent $\mathfrak{T}$-group can be insoluble.

It is well-known that every nilpotent group of class c belongs to $\mathfrak{B}_c$; on the other hand, soluble and even polycyclic groups need not be $\mathfrak{B}$-groups. A classification of finitely generated soluble $\mathfrak{B}$-groups is obtained in Section 10.5 (Part 2).

We mention two more classes of $\mathfrak{B}$-groups. *Every group which satisfies Min-sn has bounded subnormal indices*; see the Corollary to Theorem 5.49 later in this section. In addition, a *group in which each subnormal subgroup has a boundedly finite number of conjugates has bounded subnormal indices.* For let G be a group in which no subnormal subgroup has more than d conjugates and let $H \text{ sn } G$. If $N = N_G(H)$ and $C = \text{Core}_G N$, then $|G: N| \leqq d$ and consequently

$$|G: C| \leqq d!.$$

In a finite group of order $n > 2$ no subnormal index can exceed

$[\log_2(n-1)]$. Hence

$$s(G\colon H) \leqq [\log_2(d!-1)] + 1,$$

unless $d \leqq 2$, when $s(G\colon H) \leqq 2$. Thus $G \in \mathfrak{B}$.

Wreath products of $\mathfrak{B}$-groups are discussed in Robinson [5] where it is also shown that *an arbitrary soluble group can be embedded in a soluble $\mathfrak{B}$-group of the same derived length* and that *soluble $\mathfrak{B}_2$-groups can have arbitrary derived length*, unlike soluble $\mathfrak{T}$-groups (Theorems D and E).

(iv) Groups with the Subnormal Intersection Property

In any group the intersection of a finite set of subnormal subgroups is subnormal, but this conclusion is false for infinite sets of subnormal subgroups, as one can easily see from the infinite dihedral group. We will say that a group has the *subnormal intersection property* (s.i.p.) if the intersection of an arbitrary set of its subnormal subgroups is subnormal. Every finite group—and more generally every group satisfying Min-*sn*—has the s.i.p.; hence the s.i.p. is a finiteness condition.

A group G has the s.i.p. if and only if the series of successive normal closures of each subgroup of G becomes stationary after a finite number of steps. For let G have the s.i.p. and let $H \leqq G$. Then $H^{G,i}$ *sn* G for each finite i and therefore $H^{G,\omega}$, being the intersection of these $H^{G,i}$, is subnormal in G. If $H^{G,\omega} \lhd^r G$, then $H^{G,r} \leqq (H^{G,\omega})^{G,r} = H^{G,\omega} \leqq H^{G,r+1}$, so $H^{G,r} = H^{G,r+1} =$ etc. Conversely, suppose that each series of successive normal closures in G terminates after finitely many steps and let $\{H_\lambda\colon \lambda \in \Lambda\}$ be a set of subnormal subgroups of G with intersection I. By hypothesis there is an integer r such that $I^{G,r} = I^{G,r+1} = \cdots$; let $L = I^{G,r}$. Since $I \leqq H_\lambda$ *sn* G, it follows that $L \leqq H_\lambda$ for each $\lambda \in \Lambda$; hence $L = I$ and I *sn* G.

In a similar manner it may be shown that *a group G belongs to the class $\mathfrak{B}_i$ if and only if the length of the series of successive normal closures of any subgroup of G is at most i.* Thus every group with bounded subnormal indices has the s.i.p. The converse of this is false: a standard wreath product $A \sim B$ where A has prime order p and B is of type p^∞ has the s.i.p., but has unbounded subnormal indices (Robinson [3] or [5]).

We mention in passing groups with the *subnormal joinproperty* (s.j.p.); these are groups in which the join of every pair — and hence of every finite set—of subnormal subgroups is subnormal. Not every group has the s.j.p.—counterexamples have been given by Zassenhaus ([3], p. 235, Ex. 23), Robinson ([2], § 6) and Roseblade and Stonehewer ([1], Theorem E). On the other hand, a well-known theorem of Wielandt ([2], p. 214, (7)) asserts that *every group which satisfies Max-sn has the s.j.p.*; thus the s.j.p. is a finiteness condition. Wielandt's theorem implies that every group satisfying Max-*sn* even has the *generalized subnormal join property*: the join of an arbitrary set of subnormal subgroups is subnormal. *A group G has the generalized s.j.p. if and only if the union of an arbitrary chain of subnormal subgroups of G is subnormal* (Robinson [2], Lemma 8.1.) From this one easily proves that *a group with bounded subnormal indices has the generalized s.j.p.* Further results about the s.j.p. and generalized s.j.p. can be found in Robinson [2] and [4].

(v) Finiteness Conditions on Subnormal and Ascendant Abelian Subgroups

We denote by

Max-*snab* and Min-*snab*

the maximal and minimal conditions on subnormal abelian subgroups respectively. It is easy to show that a soluble group satisfying Max-*snab* is polycyclic and a periodic soluble group satisfying Min-*snab* is a Černikov group (and that periodicity is essential): the arguments here are those of the initial parts of the proofs of Theorems 3.31 and 3.32. More general results are to be found in papers of Baer [50] and Robinson [9].

By a $\lhd^{\alpha}\mathfrak{X}$-subgroup of a group G we mean an $\mathfrak{X}$-subgroup H such that $H \lhd^{\alpha} G$: here α is an ordinal number and $\mathfrak{X}$ is any class of groups. Recall that $\mathfrak{S}$ is the class of soluble groups.

Let α *be an ordinal* > 1. *If G is a group in which every* $\lhd^{\alpha}\mathfrak{A}$-*subgroup is finitely generated, then the soluble radical of G is polycyclic and contains all* $\lhd^{\alpha}\mathfrak{S}$-*subgroups of G.* If *G is a group which satisfies the minimal condition on periodic* $\lhd^{\alpha}\mathfrak{A}$-*subgroups, then G has a unique maximal normal periodic soluble subgroup and this is a Černikov group and contains all periodic* $\lhd^{\alpha}\mathfrak{S}$-*subgroups of G* (Robinson [9], Theorems A and A*).

From the second of these results we deduce that *if* α *is an ordinal* > 1, *the minimal condition on periodic* $\lhd^{\alpha}\mathfrak{A}$-*subgroups is equivalent to the requirement that each periodic* $\lhd^{\alpha}\mathfrak{A}$-*subgroup satisfy Min.* The corresponding statement for the maximal condition on $\lhd^{\alpha}\mathfrak{A}$-subgroups is true if α is infinite but false when $\alpha = 2$. (Robinson [9], Theorem F). However, Max-*snab* is equivalent to every subnormal abelian subgroup being finitely generated: the corresponding statement for Min-*snab* is trivially true.

A subsoluble group satisfying Max-snab is polycyclic and a periodic subsoluble group satisfying Min-snab is a soluble Černikov group (Baer [50], Hauptsätze 5.1 and 6.1, Robinson [9], Theorem E). These results follow easily from the theorems quoted in the last paragraph but one. For further details the reader is referred to the papers of Baer and Robinson already mentioned.

Subnormal Simple Subgroups

We discuss next the special role played by the subnormal simple subgroups of a group: notice that these are precisely the *minimal subnormal subgroups*.

Theorem 5.42 (Wielandt [4]). Let H and K be subnormal subgroups of a group and let $H \cap K = 1$. If H is a non-abelian simple group, then $[H, K] = 1$.

Proof. Let $J = \langle H, K \rangle$ and $s = s(J: H)$. Suppose first of all that $s \leqq 1$; then $H \lhd J$ and $[H, K] \lhd H$, which implies that either $[H, K] = 1$ or $[H, K] = H$. In the latter event a normal subgroup of J which contains K must also contain H and therefore coincides with J. Since $K \text{ sn } J$, it follows that $K = J$ and $H = H \cap J = 1$, which is not so.

Let $s > 1$ and let us assume that the theorem is true for $s - 1$. Now H is not normal in J, so there is a $k \in K$ such that $H \neq H^k$. Since $H \cap H^k \text{ sn } J$ and both H and H^k are simple, $H \cap H^k = 1$. Also $s(H^J: H) = s - 1$; thus $[H, H^k] = 1$ by induction on s. Let h and h_1 belong to H; then

$$1 = [h, h_1^k] = [h, h_1[h_1, k]] = [h, [h_1, k]]\,[h, h_1]^{[h_1, k]}.$$

Consequently $H' \leqq [H, K]$. But $H = H'$ because H is simple and non-abelian; hence $H \leqq [H, K]$. This leads as before to the contradiction $K = J$. ∎

Considerable attention has been paid to pairs of subnormal subgroups with trivial intersection. One may ask, for example, when such subgroups commute elementwise or when one normalizes the other; for these and related questions see Wielandt [2], [3] and Roseblade [4], [7].

The Wielandt Subgroup

If G is an arbitrary group, we define *the Wielandt subgroup* of G

$$\omega(G)$$

to be the intersection of all the normalizers of subnormal subgroups of G (Wielandt [4]). This is analogous to the *kern* or *norm* of G, i.e. the intersection of all the normalizers of arbitrary subgroups of G, which had previously been introduced by Baer ([2], p. 254).

Clearly $\omega(G)$ is always a $\mathfrak{T}$-group and indeed G is a $\mathfrak{T}$-group if and only if $G = \omega(G)$. Theorem 5.42 implies that *all subnormal non-abelian simple subgroups are contained in the Wielandt subgroup*. Here is a somewhat similar result.

Theorem 5.43 (Wielandt [4]). A minimal normal subgroup which satisfies Min-n lies in the Wielandt subgroup.

Proof. Let N be a minimal normal subgroup of a group G and assume that N satisfies Min-n. Then Corollary 2 to Lemma 5.23 implies that N even satisfies Min-sn. Let H be any subnormal subgroup of G and put $s = s(G: H)$. We have to prove that N normalizes H; if $s \leqq 1$, this is obvious, so let $s > 1$. Since N is minimal normal, either $H^G \cap N = 1$ — in which case $[H, N] \leqq H^G \cap N = 1$ and, *a fortiori*, $H^N = N$— or $N \leqq H^G$. Suppose that we have the latter situation. Since N satisfies Min-sn, there is a minimal normal subgroup of H^G contained in N, say M. Clearly, each conjugate of M in G is a minimal normal subgroup of H^G and satifies Min-sn. Now $s(H^G: H) = s - 1$, so by induction on s it follows that each conjugate of M normalizes H. By minimality of N we have $N = M^G$, so N normalizes H. □

Corollary. If G satisfies Min-sn, the socle of G is contained in the Wielandt subgroup and if G is non-trivial, the Wielandt subgroup is non-trivial.

In general a minimal normal subgroup need not lie in the Wielandt subgroup: an example which shows this is on p. 81 (Part 2). Also it is not hard to find non-trivial groups with trivial Wielandt subgroup: the infinite dihedral group is of this type.

Direct Products of Simple Groups

Lemma 5.44. The subgroup generated by a set of subnormal non-abelian simple subgroups of a group is the direct product of certain members of the set.

The proof is of exactly the same form as that of Lemma 5.23, except that we require Theorem 5.42 in the last stage.

Direct products of simple groups may be characterized as follows.

Theorem 5.45. (i) A subnormal subgroup of a direct product of simple groups is a direct factor and is also a direct product of simple groups.

(ii) A group such that every proper normal subgroup is contained in a proper direct factor is a direct product of simple groups.

The first part of this theorem is due to Remak ([1], Satz 8); the second part is due to Head [1] and, in a special case, to Wiegold [3].

Proof.

(i) Let G be a direct product of simple groups. It is clearly sufficient to prove the result for a normal subgroup N of G.

For the moment let us suppose that G is the direct product of *non-abelian* simple groups G_α, $\alpha \in A$. If $N \cap G_\alpha = 1$ for all $\alpha \in A$, then $[N, G_\alpha] = 1$ and $N \leq \zeta(G)$. But $\zeta(G) = 1$ since each G_α has trivial centre; thus $N = 1$. Now for each $\alpha \in A$ either $N \cap G_\alpha = 1$ or $G_\alpha \leq N$; let P be the product of all the G_α that are contained in N. Then $N/P \lhd G/P$ and $N = P$ by our first conclusion. *In this case N is actually the direct product of certain of the G_α*; hence N is direct factor of G.

Now let G be any direct product of simple groups; then we can write

$$G = G_0 \times G_1$$

where G_0 is a direct product of non-abelian simple groups and G_1 is a direct product of cyclic groups with prime orders. By the previous paragraph $N \cap G_0$ is a direct factor of G_0, and obviously $N \cap G_1$ is a direct factor of G_1; hence $P = (N \cap G_0) \times (N \cap G_1)$ is a direct factor of G. Finally, N/P is a homomorphic image of a normal subgroup of G_0, so it cannot be abelian and non-trivial; yet it is also a homomorphic image of a subgroup of G_1. Therefore $N = P$.

(ii) Assume that G has the property specified in (ii) of the theorem and let P be the product of all the normal simple subgroups of G. If $P = G$, then we are finished in view of Lemma 5.23. Suppose that $P \neq G$ and let $x \in G \backslash P$. By Zorn's Lemma there is a normal subgroup M of G which is maximal subject to the requirements $P \leq M$ and $x \notin M$. Suppose that D is a proper direct factor of G containing M: let us write $G = D \times E$. Since $E \neq 1$, the maximality of M shows that $x \in M \times E$. Since $D \cap (M \times E) = M$, it follows that $x \notin D$. Therefore $D = M$ by maximality of M. Thus the only direct factor of G which properly contains M is G. Now suppose that N is a proper non-trivial normal subgroup of E; then $M \times N$ is a proper normal subgroup of G and so lies in a proper direct factor of G. This contradicts the conclusion of the previous sentence, so E is simple and therefore $E \leq P \leq M$, our final contradiction. □

Corollary. If H is a subnormal non-abelian simple subgroup of a group G, then H^G is a minimal normal subgroup of G. Also $H \lhd^2 G$, and even an ascendant non-abelian simple subgroup of G is subnormal in G with subnormal index ≤ 2.

Proof. By Lemma 5.44 the group H^G is a direct product of conjugates of H and by the first part of the proof of Theorem 5.45 a non-trivial

G-admissible subgroup of H^G contains a conjugate of H and hence coincides with H^G. The remaining statements are clear. ▯

The information just obtained may be used to get criteria for a group satisfying Min-*sn* or Max-*sn* to be Černikov or polycyclic-by-finite respectively.

Theorem 5.46. Let G be a group which satisfies Min-*sn* (Max-*sn*). Then G is a Černikov group (a polycyclic-by-finite group) if and only if G has an ascending series with finite or locally nilpotent factors.

Proof. Let G be a non-trivial group with the property Min-*sn* and assume that G also satisfies the condition of the theorem. Suppose that R, the Hirsch-Plotkin radical of G, is non-trivial; then R is a locally nilpotent group satisfying Min-*sn* and Corollary 2 to Theorem 5.27 shows that R is a Černikov group. Hence R contains a non-trivial G-admissible finite subgroup. Suppose on the other hand that $R = 1$; then by Theorem 2.31 the group G has no non-trivial ascendant locally nilpotent subgroups. In this case G must have a non-trivial ascendant finite subgroup and the latter contains a subnormal non-abelian simple subgroup H. Lemma 5.44 shows that H^G is a direct product of conjugates of H. Since H^G satisfies Min-*sn* and H is finite, we conclude that H^G is finite. Thus in either case G contains a non-trivial normal finite subgroup.

Since homomorphic images inherit the hypotheses on G, the argument just given shows that G is hyperfinite. That G is a Černikov group now follows by Corollary 2 to Theorem 5.21: the converse assertion is trivially true. For Max-*sn* the proof is similar. ▯

Corollary. A group which satisfies Min-*sn* is a Černikov group if and only if each of its subnormal composition factors is finite.

The first part of Theorem 5.46 may be found in Amberg [3]. We mention in passing two analogous results (cf. Plotkin [10], Baer [31], Satz B, and Baer [55], Proposition II.5). *A group satisfying Max-ab (Min-ab) is polycyclic-by-finite (Černikov) if and only if it has an ascending series whose factors are locally finite or locally nilpotent (finite or locally soluble).* This may be proved by applying Theorems 3.31, 3.32, 3.43, 3.45 and 4.39. The proof proceeds thus: first factor out the maximal normal radical subgroup, observing that this is polycyclic (Černikov). Then factor out the finite radical, which turns out to be finite. The resulting group is easily seen to be trivial.

The Subnormal Socle of a Group

We define the *subnormal socle* of a group G to be the subgroup S generated by all the minimal subnormal (i.e. subnormal simple) subgroups of G.

Let S_0 and S_1 denote respectively the subgroups generated by the non-abelian and the abelian, minimal subnormal subgroups of G. Then Lemma 5.44 and the proof of Theorem 5.45 (i) show that S_0 is the direct product of *all* the non-abelian minimal subnormal subgroups of G. Also S_1 is a periodic Baer group and, of course, $S = S_0 S_1$. Clearly S_0 has no normal Baer subgroups other than the identity subgroup, so $S_0 \cap S_1 = 1$ and

$$S = S_0 \times S_1. \tag{8}$$

Moreover, an arbitrary minimal subnormal subgroup of G is either a direct factor of S_0 or else is abelian and is contained in S_1 since S_1 is the Baer radical of S.

Although in general the socle and the subnormal socle are incomparable, they do have S_0 in common (Corollary to Theorem 5.45); also if the socle has Min-n, it is contained in the subnormal socle (Corollary 2 to Lemma 5.23).

Lemma 5.47. Let S be the subnormal socle of a group G; then the following statements are equivalent.

(i) G has a finite number of minimal subnormal subgroups.

(ii) S is the direct product of a finite nilpotent group and a finite number of non-abelian simple groups.

(iii) S satisfies Min-*sn*.

(iv) S satisfies Min-*n*.

Proof. We write $S = S_0 \times S_1$ with the notation of equation (8). Let G satisfy (i); then S_0 is the direct product of finitely many non-abelian simple groups. Also S_1 is a finitely generated periodic Baer group, so it is finite and nilpotent. Thus (i) implies (ii). That (ii) implies (iii) and (iii) implies (iv) is trivial. Finally let G satisfy (iv); then S_0 is the direct product of finitely many non-abelian simple groups and these are the only non-abelian minimal subnormal subgroups of G, by the proof of Theorem 5.45 (i). Also S_1, being a Baer group with the property Min-*n*, is nilpotent (p. 155); now S_1/S_1' is generated by elements of prime order and satisfies Min and this implies that S_1/S_1' is finite. The finiteness of S_1 is now a consequence of the Corollary to Theorem 2.26. Every abelian minimal subnormal subgroup of G lies in S_1; thus G satisfies (i). □

Characterizations of Groups which Satisfy Min-*sn*

Our principal objective is the following theorem.

Theorem 5.48 (Robinson [4], Theorem 3.1). The following properties of a group G are equivalent.

(i) G satisfies Min-*sn*.

(ii) To each proper subnormal subgroup H of G there corresponds at least one but only finitely many subnormal subgroups of G which are minimal with respect to properly containing H.

(iii) The subnormal socle of each non-trivial homomorphic image of G is non-trivial and satisfies Min-*n*.

First of all we will establish

Theorem 5.49 (Roseblade [3]; Robinson [4], Lemma 3.2). If G is a group satisfying Min-*sn*, the Wielandt subgroup of G has finite index.

Proof. Let R be the finite residual of G; to show that the Wielandt subgroup of G has finite index it will be enough to prove that R normalizes an arbitrary subnormal subgroup H of G. Suppose that H is minimal subject to the requirements $H \text{ sn } G$ and $H^R \neq H$, and let P be the subgroup generated by all the proper subnormal subgroups of H; then $P^R = P$ by minimality of H. Hence P is a proper normal subgroup of H and clearly H/P is simple. Now $P \lhd HR$ and $HR \text{ sn } G$, so HR/P satisfies Min-*sn* and therefore contains only a finite number of minimal subnormal subgroups. If $x \in R$, then H^x/P is a minimal subnormal subgroup of HR/P and consequently H has only a finite number of conjugates in R. But R has no proper subgroups of finite index, so $H^R = H$ and the theorem is established. ▯

From this and Theorem 5.43 we deduce

Corollary. Let the group G satisfy Min-*sn*. Then the upper Wielandt series reaches G after a finite number of steps and G has bounded subnormal indices.

Proof of Theorem 5.48. Let the group G satisfy Min-*sn*: we show first that G satisfies condition (ii). Let H be a proper subnormal subgroup of G. Then by Min-*sn* there exist subnormal subgroups of G that are minimal with respect to properly containing H; suppose that $\{K_1, K_2, \ldots\}$ is a countably infinite set of distinct subgroups of this type. By minimality $H \lhd K_i$ and K_i/H is simple for each i. Let $J = \langle K_1, K_2, \ldots \rangle$; then $H \lhd J$ and the K_i/H are minimal subnormal subgroup of J/H. Denote by R the finite residual of G; then R normalizes each K_i, by Theorem 5.49, so $J \lhd JR$. Since G/R is finite, the subnormal subgroups K_iR/R generate a subnormal subgroup of G/R, namely JR/R (Wielandt [2], p. 214, (7)). Thus $J \lhd JR \text{ sn } G$ and $J \text{ sn } G$, from which we infer that J, and hence J/H, satisfies Min-*sn*. But J/H has in the K_i/H an infinite set of minimal subnormal subgroups, which contradicts Lemma 5.47. Thus (i) implies (ii). That (ii) implies (iii) is an immediate consequence of Lemma 5.47.

It remains to show that a group G which has the property (iii) must satisfy Min-*sn*. Let $G \neq 1$ we begin by forming the ***upper subnormal socle series*** of G; this is the ascending normal series $\{S_\alpha\}$ defined by

$$S_{\alpha+1}/S_\alpha = \text{the subnormal socle of } G/S_\alpha \ .$$

and

$$S_0 = 1, \quad S_\lambda = \bigcup_{\beta<\lambda} S_\beta$$

where α is an ordinal and λ a limit ordinal. Clearly $G = S_\beta$ for some ordinal β.

The structure of S_1 is described by Lemma 5.47; because of this and the Corollary to Theorem 5.45, the subgroup S_1 contains a minimal normal subgroup of G, and the socle of G is non-trivial. Let N be any minimal normal subgroup of G; then there is a first ordinal α, clearly not a limit ordinal, such that $N \cap S_\alpha \neq 1$. Thus $N \cap S_{\alpha-1} = 1$ and

$$N \simeq NS_{\alpha-1}/S_{\alpha-1} \lhd S_\alpha/S_{\alpha-1}.$$

Now by hypothesis $S_\alpha/S_{\alpha-1}$ satisfies Min-*n* and hence Min-*sn* (by Lemma 5.47), so N contains a minimal subnormal subgroup and consequently $N \cap S_1 \neq 1$. Hence $\alpha = 1$ and $N \leqq S_1$. It follows that the socle of G is contained in the subnormal socle S_1; therefore the socle of G satisfies Min-*sn*. Since this conclusion applies equally to every non-trivial homomorphic image of G, Theorem 5.24 may now be invoked to show that G satisfies Min-*n*.

Let R be the finite residual of G; then G/R is finite and R satisfies Min-*n* by Theorem 5.21. It will be enough therefore if we can prove that every subnormal subgroup of R is normal in R.

We can refine the series $\{S_\alpha\}$ to an ascending series whose terms are subnormal in G and whose factors are simple: here we are using Lemma 5.47. Let H be a subnormal subgroup of R; if we intersect H with the refined series in G we obtain (after deleting any repetitions) an ascending subnormal series in H with simple factors: the length of this series will be called the *height* of H. Let α be the first ordinal such that there exists a non-normal subnormal subgroup H of R with height α. Then it is clear that α cannot be a limit ordinal. Hence there exists a normal subgroup K of H such that the height of K is $\alpha - 1$ and H/K is a simple group. If S is any subnormal subgroup of K, then the height of S is $\leqq \alpha - 1$ and consequently $S \lhd R$. Since R satisfies Min-*n*, we conclude that K satisfies Min-*sn*. Now $K \lhd R$ and $|G\colon R|$ is finite, so K has only a finite number of conjugates in G; each of these is a normal subgroup of R and satisfies Min-*sn*. Hence $N = K^G$ satisfies Min-*sn*, because the latter property, being ***P*** and ***H***-closed, is $\mathbf{N}_0$-closed. Let T/K denote the subnormal socle of R/K; then $T \lhd R$, so $T \cap N \lhd N$ and $T \cap N$ satisfies Min-*sn*. Also $T/T \cap N \simeq TN/N$ and the latter, being a homomorphic image of T/K, is generated by minimal subnormal subgroups of G/N; thus TN/N is a subnormal subgroup of the subnormal socle of G/N, which, by hypothesis, satisfies Min-*n* and hence Min-*sn*. Thus TN/N—and hence T—satisfies Min-*sn*. Let $x \in R$; then $K \lhd H^x$ and H^x/K is a minimal subnormal subgroup of R/K and thus lies in T/K. Since T/K satisfies Min-*sn*, Lemma 5.47 implies that H can have only finitely many conjugates in R; consequently $H^R = H$, contrary to our assumption. ▯

Corollary (Wielandt [2], p. 219). A group has a subnormal composition series of finite length if and only if it contains only finitely many subnormal subgroups.

Proof. Assume that G is a group with a subnormal composition series of finite length. By the Schreier Refinement Theorem (Kuroš [9], vol. 1, p. 111) every subnormal subgroup appears in some subnormal composition series, thus it is enough if we can prove that there are only finitely many such series in G. The Jordan-Hölder Theorem (Kuroš [9], vol. 1, p. 112) asserts that all subnormal composition series of G have the same finite length; also G has the property (ii) of Theorem 5.48, so a subnormal composition series can be formed in just a finite number of ways. ▯

We shall apply Theorem 5.48 to obtain further characterizations of the property Min-*sn*.

Theorem 5.49.1. The following group theoretical properties are equivalent:

(i) the minimal condition on subnormal subgroups,

(ii) the minimal condition on subnormal subgroups with subnormal index at most 2,

(iii) each normal subgroup satisfies the minimal condition on normal subgroups,

(iv) each characteristic subgroup satisfies the minimal condition on normal subgroups.

Proof. The implications (i) → (ii) → (iii) → (iv) are obvious. Let the non-trivial group G have property (iv). We shall prove that each non-trivial factor group of G has non-trivial subnormal socle satisfying Min-*n*, which will show that G satisfies Min-*sn* by Theorem 5.48.

Since G itself has Min-*n*, its upper socle series terminates at G. Moreover, each term of this series is characteristic in G and hence satisfies Min-*n*. Thus each factor of the upper socle series of G satisfies Min-*n*—and even Min-*sn* by Corollary 2 to Lemma 5.23. A chief factor of G is a homomorphic image of a normal subgroup of some factor of the upper socle series of G. Consequently every chief factor of G has Min-*sn*.

Now let N be a proper normal subgroup of G. Since G satisfies Min-*n*, so does G/N and consequently G/N has a minimal normal subgroup; this, of course, is a chief factor of G, so it satisfies Min-*sn* and, by Corollary 2 to Lemma 5.23, it lies in the subnormal socle of G/N, which is therefore non-trivial.

Let H/N be a subnormal non-abelian simple subgroup of G/N. Then H^G/N is a minimal normal subgroup of G/N (Corollary to Theorem 5.45). Consequently the subnormal non-abelian simple subgroups of G/N generate a normal subgroup K/N contained in the socle of G/N. The latter satisfies Min-*sn* since it is the direct product of finitely many minimal normal subgroups of G/N each of which satisfies Min-*sn*. Hence K/N satisfies Min-*sn*. The subnormal abelian simple subgroups of G/N all lie in its Baer radical B/N. Since G has Min-*n*, the subgroups N and B are terms of some ascending chief series of G. Now a chief factor of G satisfies Min-*n*—if in addition it is a Baer group, it is a Černikov group (Corollary 2 to Theorem 5.27) and hence must be finite. It follows that B/N lies in the FC-hypercentre of G/N. Theorem 5.22 (i) implies that B/N is a Černikov group. The subnormal socle of G/N lies in the product of the normal subgroups K/N and B/N, both of which have Min-*sn*. Hence the subnormal socle of G/N satisfies Min-*sn*. ▯

We remark that an unpublished example of R. S. Dark shows that there exists a group which satisfies the maximal condition on subnormal subgroups with subnormal index $\leqq i$ for each $i \geqq 0$, but which does not satisfy Max-*sn*. So there is no analogue of Theorem 5.49.1 for the maximal condition.

Further Results

We conclude this chapter by mentioning without proof some related results. *There exist groups satisfying Min-sn which have no proper subgroups of finite index and which have a unique ascending chief series of arbitrary ordinal type* (Robinson [4], Theorem 5.1); by Theorem 5.49 such groups are $\mathfrak{Z}$-groups. In [3] Roseblade has shown that *in an arbitrary group a subnormal subgroup with the property Min-sn and without proper subgroups of finite index normalizes every subnormal subgroup with the property Min-sn*; this generalization of Theorem 5.49 has also been proved by Robinson ([4], Lemma 4.3) and has been slightly generalized by Kegel ([3], Satz 1.7).

As a consequence of this theorem of Roseblade one may prove that *an* $\mathbf{N}_0$*-closed subclass of Min-sn is a subnormal coalition class* (Roseblade [3], Robinson [4], Theorem 4.1) and that *a* **P**- *and* **H**-*closed subclass of*

Min-sn is an ascendant coalition class (Robinson [4], Theorem 4.2). Roseblade has also shown that ***an*** $\mathbf{N}_0$*-closed subclass of Max-sn is a subnormal coalition class* [5].

Analogues of some of the results in this section—for example Theorems 5.49 and 5.49.1 — have been proved for Lie algebras over fields of characteristic 0 by I. N. Stewart [1].

Bibliography

The following papers are referred to in the text. This list is an extract from the complete bibliography which appears at the end of Part 2. Russian language papers are marked with an asterisk and, as is customary, their titles have been translated into English. Where an English translation of such a work exists, this is given immediately after the original.

Abramovskiĭ, I. N.

*1. Locally generalized Hamiltonian groups. Sibirsk. Mat. Ž. **7**, 481–485 (1966) = Siberian Math. J. **7**, 391–393 (1966).

*2. The structure of locally generalized Hamiltonian groups. Leningrad Gos. Ped. Inst. Učen. Zap. **302**, 43–49 (1967).

Abramovskiĭ, I. N., Kargapolov, M. I.

*1. Finite groups with the property of transitivity for normal subgroups. Uspehi Mat. Nauk. **13**, 232–243 (1958).

Ado, I. D.

*1. On nilpotent algebras and p-groups. Dokl. Akad. Nauk. SSSR **40**, 299–301 (1943).

*2. On subgroups of the countable symmetric group. Dokl. Akad. Nauk. SSSR **50**, 15–18 (1945).

*3. On locally finite p-groups with the minimal condition for normal subgroups. Dokl. Akad. Nauk. **54**, 471–473 (1946).

Alperin, J. L.

1. Groups with finitely many automorphisms. Pacific J. Math. **12**, 1–5 (1962).

Alperin, J. L., Brauer, R., Gorenstein, D.

1. Finite groups with quasi-dihedral and wreathed Sylow 2-subgroups. Trans. Amer. Math. Soc. **151**, 1–261 (1970).

Amberg, B.

3. Gruppen mit Minimalbedingung für Subnormalteiler. Arch. Math. (Basel) **19**, 348–358 (1968).

Amitsur, S.

1. A general theory of radicals. Amer. J. Math. **74**, 774–786 (1952); **76**, 100–125 (1954); **76**, 126–136 (1954).

Auslander, L.

2. The automorphism group of a polycyclic group. Ann. of Math. (2) **89**, 314–322 (1970).

Auslander, L., Baumslag, G.

1. Automorphism groups of finitely generated nilpotent groups. Bull. Amer. Math. Soc. **73**, 716–717 (1967).

Bačurin, G. F.

*7. On multipliers of torsion-free nilpotent groups. Mat. Zametki **3**, 541–544 (1969) = Math. Notes **3**, 325–327 (1969).

Baer, R.

2. Der Kern, eine charakteristische Untergruppe. Composito Math. **1**, 254–283 (1934).
3. Die Kompositionsreihe der Gruppe aller eineindeutigen Abbildungen einer unendlichen Menge auf sich. Studia Math. **5**, 15–17 (1934).
7. Nilpotent groups and their generalizations. Trans. Amer. Math. Soc. **47**, 393–434 (1940).
10. Groups without proper isomorphic quotient groups. Bull. Amer. Math. Soc. **50**, 267–278 (1944).
11. Representations of groups as quotient groups. Trans. Amer. Math. Soc. **58**, 295–419 (1945).
12. Finiteness properties of groups. Duke Math. J. **15**, 1021–1032 (1948).
13. Groups with descending chain condition for normal subgroups. Duke Math. J. **16**, 1–22 (1949).
14. Endlichkeitskriterien für Kommutatorgruppen. Math. Ann. **124**, 161–177 (1952).
16. The hypercenter of a group. Acta Math. **89**, 165–208 (1953).
20. Nil-Gruppen. Math. Z. **62**, 402–437 (1955).
21. Supersoluble groups. Proc. Amer. Math. Soc. **6**, 16–32 (1955).
23. Finite extensions of abelian groups with minimum condition. Trans. Amer. Math. Soc. **79**, 521–540 (1955).
24. Auflösbare Gruppen mit Maximalbedingung. Math. Ann. **129**, 139–173 (1955).
26. Noethersche Gruppen. Math. Z. **66**, 269–288 (1956/1957).
27. Lokal Noethersche Gruppen. Math. Z. **66**, 341–363 (1956/1957).
31. Kriterien für die Endlichkeit von Gruppen. Jahresber. Deutsch. Math. Verein. **63**, 53–77 (1960).
32. Abzählbar erkennbare gruppentheoretische Eigenschaften. Math. Z. **79**, 344–363 (1962).
34. Gruppen mit Minimalbedingung. Math. Ann. **150**, 1–44 (1963).
35. Irreducible groups of automorphisms of abelian groups. Pacific J. Math. **14**, 385–406 (1964).
39. Groups with minimum condition. Acta Arith. **9**, 117–132 (1964).
41. Noethersche Gruppen II. Math. Ann. **165**, 163–180 (1966).
42. Endlich definierbare gruppentheoretische Funktionen. Math. Z. **87**, 163–213 (1965).
43. Group theoretical properties and functions. Colloq. Math. **14**, 285–328 (1966).
48. Soluble artinian groups. Canad. J. Math. **19**, 904–923 (1967).
50. Auflösbare, artinsche, noethersche Gruppen. Math. Ann. **168**, 325–363 (1967).
51. Polyminimaxgruppen. Math. Ann. **175**, 1–43 (1968).

52. Gruppen mit abzählbaren Automorphismengruppen. Hamburg Math. Einzelschr. **2** (1969).
53. Lokal endliche-auflösbare Gruppen mit endlichen Sylowuntergruppen. J. reine angew. Math. **239/240**, 109–144 (1969).
55. The determination of groups by their groups of automorphisms. Studies in Pure Mathematics (1971).

Balcerzyk, S.
1. On classes of abelian groups. Bull. Acad. Polon. Sci. Ser. Sci. Math. Astronom. Phys. **9**, 327–329 (1961).
2. On classes of abelian groups. Fund. Math. **51**, 149–179 (1962).

Bass, H., Lazard, M., Serre, J.-P.
1. Sous-groupes d'indice fini dans SL(n, Z). Bull. Amer. Math. Soc. **70**, 385–392 (1964).

Baumslag, G.
9. Wreath products and finitely presented groups. Math. Z. **75**, 22–28 (1961).
10. A non-hopfian group. Bull. Amer. Math. Soc. **68**, 196–198 (1962).
13. On abelian hopfian groups I. Math. Z. **78**, 53–54 (1962).
20. Groups with one defining relator. J. Austral. Math. Soc. **4**, 385–392 (1964).

Baumslag, G., Solitar, D.
1. Some two-generator one-relator non-Hopfian groups. Bull. Amer. Math. Soc. **68**, 199–201 (1962).

Bender, H.
1. Finite groups having a strongly embedded subgroup. Theory of finite groups, ed. R. Brauer and C. H. Sah: Benjamin 1969, pp. 21–24.

Berlinkov, M. L.
*1. On the lattice of subgroups of a group with finite layers. Uspehi Mat. Nauk. **12**, 267–271 (1957).

Berman, S. D., Lyubimov, V. V.
*1. Groups allowing arbitrary permutations of the factors of their composition series. Uspehi Mat. Nauk. **12**, 181–183 (1957).

Best, E., Taussky, O.
1. A class of groups. Proc. Roy. Irish Acad. Sect. A **47**, 55–62 (1942).

Betten, A.
1. Hinreichende Kriterien für die Hyperzentralität einer Gruppe. Arch. Math. (Basel) **20**, 471–480 (1970).

Birkhoff, G.
1. Transfinite subgroup series. Bull. Amer. Math. Soc. **40**, 847–850 (1934).
2. The structure of abstract algebras. Proc. Cambridge Philos. Soc. **31**, 433–454 (1935).

Blackburn, N.
2. On a special class of p-groups. Acta Math. **100**, 45–92 (1958).

Bowers, J. F.
1. On composition series of polycyclic groups. J. London Math. Soc. **35**, 433–444 (1960).

Brauer, R.
1. Some applications of the theory of blocks of characters II. J. Algebra **1**, 307–334 (1964).

Brenner, J. L.
1. Quelques groupes libres de matrices. C. R. Acad. Sci. Paris **241**, 1689–1691 (1955).
2. The linear homogeneous group III. Ann. of Math. (2) **71**, 210–223 (1960).

Bride, I. M.
1. Second nilpotent BFC-groups. J. Austral. Math. Soc. **11**, 9–18 (1970).

Burnside, W.
5. On criteria for the finiteness of the order of a group of linear substitutions. Proc. London Math. Soc. (2) **3**, 435–440 (1905).

Calenko, M. S.
*1. Some remarks on infinite simple groups. Sibirsk. Mat. Ž. **4**, 227–231 (1963).

Camm, R.
1. Simple free products. J. London Math. Soc. **28**, 66–76 (1953).

Čan Van Hao
*1. On semi-simple classes of groups. Sibirsk. Mat. Ž. **3**, 943–949 (1962).

Čarin, V. S.
*1. A remark on the minimal condition for subgroups. Dokl. Akad. Nauk. SSSR **66**, 575–576 (1949).
*3. On the theory of locally nilpotent groups. Mat. Sb. **29**, 433–454 (1951) = Amer. Math. Soc. Translations (2) **15**, 33–54 (1960).
*4. On the minimal condition for normal subgroups of locally soluble groups. Mat. Sb. **33**, 27–36 (1953).
*6. On groups of automorphisms of nilpotent groups. Ukrain. Mat. Ž. **6**, 295–304 (1954).
*9. On groups possessing soluble ascending invariant series. Mat. Sb. **41**, 297–316 (1957).

Černikov, S. N.
*4. Infinite locally soluble groups. Mat. Sb. **7**, 35–64 (1940).
*12. Infinite groups with finite layers. Mat. Sb. **22**, 101–133 (1948) = Amer. Math. Soc. Translations (1) **56**, 51–102 (1951.
*16. On the theory of locally soluble groups with the minimal condition for subgroups Dokl. Akad. Nauk. SSSR **65**, 21–24 (1949).
*19. Periodic ZA-extensions of complete groups. Mat. Sb. **27**, 117–128 (1950).
*20. On special p-groups. Mat. Sb. **27**, 185–200 (1950).
*21. On the centralizer of a complete abelian normal subgroup of an infinite periodic group. Dokl. Akad. Nauk. SSSR **72**, 243–246 (1950).
*22. On locally soluble groups which satisfy the minimal condition for subgroups. Mat. Sb. **28**, 119–129 (1951).
*23. On groups with finite classes of conjugate elements. Dokl. Akad. Nauk. SSSR **114**, 1177–1179 (1957).
*24. On the structure of groups with finite classes of conjugate elements. Dokl. Akad. Nauk. SSSR **115**, 60–63 (1957).
*25. On layer-finite groups. Mat. Sb. **45**, 415–416 (1958).
*26. Finiteness conditions in the general theory of groups. Uspehi Mat. Nauk. **14**, 45–96 (1959) = Amer. Math. Soc. Translations (2) **84**, 1–67 (1969) (with a supplement by the author).
29. Endlichkeitsbedingungen in der Gruppentheorie, Math. Forschungsberichte XX, Berlin: VEB Deutscher Verlag der Wissenschaften, 1963; (German translation of [26] with a supplement by the author).

*34. Periodic groups of automorphisms of extremal groups. Math. Zametki 4, 91–96 (1968) = Math. Notes 4, 543–545 (1968).
*35. Infinite non-abelian groups with the minimal condition for non-normal abelian subgroups. Dokl. Akad. Nauk. SSSR 184 (1969), 786–789 (1969) = Soviet Math. Dokl. 10, 172–175 (1969).

Chehata, S.
1. An algebraically simple ordered group. Proc. London Math. Soc. (3) 2, 183–197 (1952).

Clapham, C. R. J.
1. Finitely presented groups with word problems of arbitrary degrees of insolubility. Proc. London Math. Soc. (3) 14, 633–676 (1964).
2. An embedding theorem for finitely generated groups. Proc. London Math. Soc. (3) 17, 419–430 (1967).

Clowes, J. S., Hirsch, K. A.
1. Simple groups of infinite matrices. Math. Z. 58, 1–3 (1953).

Curtis, C., Reiner, I.
1. Representation theory of finite groups and associative algebras. New York Interscience 1962.

Dark, R. S., Rhemtulla, A. H.
1. On $\mathbf{R}_0$-closed classes and finitely generated groups. Canad. J. Math. 22, 176–184 (1970).

Dickson, S. E.
1. On torsion classes of abelian groups. J. Math. Soc. Japan. 17, 30–35 (1965).

Dietzmann (Dicman), A. P.
*1. On p-groups. Dokl. Akad. Nauk. SSSR 15, 71–76 (1937).

Dixmier, S.
1. Exposants des quotients des suites centrales descendantes et ascendantes d'un groupe. C. R. Akad. Sci. Paris 259, 2751–2753 (1964).

Dixon, J. D.
3. The Fitting subgroup of a linear solvable group. J. Austral. Math. Soc. 7, 417–424 (1967).
4. The solvable length of a solvable linear group. Math. Z. 107, 151–158 (1968).

Dlab, V.
1. On cyclic groups. Czechoslovak Math. J. 10, 244–254 (1960).
2. A note on powers of a group. Acta Sci. Math. (Szeged) 25, 177–178 (1964).
3. A remark on a paper of Gh. Pic. Czechoslovak Math. J. 17, 467–468 (1967).

Duguid, A. M.
1. A class of hyper FC-groups. Pacific J. Math. 10, 117–120 (1960).

Duguid, A. M., McLain, D. H.
1. FC-nilpotent and FC-soluble groups. Proc. Cambridge Philos. Soc. 52, 391–398 (1956).

Erdös, J.
1. The theory of groups with finite classes of conjugate elements. Acta Math. Acad. Sci. Hungar. 5, 45–58 (1954).

Eremin, I. I.
*1. Groups with finite classes of conjugate abelian subgroups. Mat. Sb. 47, 45–54 (1959).
*2. Groups with finite classes of conjugate infinite subgroups. Perm Gos. Univ. Učen. Zap. 17, 13–14 (1960).

*3. On central extensions by means of thin layer-finite groups. Izv. Vysš. Učebn. Zaved. Matematika **15**, 93–95 (1960).

*4. Groups with finite classes of conjugate subgroups with a given property. Dokl. Akad. Nauk. SSSR **137**, 772–773 (1961) = Soviet Math. Dokl. **2**, 337–338 (1961).

Fedorov, Yu. G.

*1. On infinite groups every non-trivial subgroup of which has finite index. Uspehi Mat. Nauk. **6**, 187–189 (1951).

Feit, W., Thompson, J. G.

1. Solvability of groups of odd order. Pacific J. Math. **13**, 775–1029 (1963).

Fitting, H.

1. Beiträge zur Theorie der Gruppen endlicher Ordnung. Jahresber. Deutsch. Math. Verein. **48**, 77–141 (1938).

Fuchs, L.

1. On groups with finite classes of isomorphic subgroups. Publ. Math. Debrecen **3**, 243–252 (1954).

2. The existence of indecomposable abelian groups of arbitrary power. Acta Math. Acad. Sci. Hungar. **10**, 453–457 (1959).

3. Abelian groups. Oxford: Pergamon 1960.

Gaschütz, W.

1. Gruppen in denen das Normalteilersein transitiv ist. J. reine angew. Math. **198**, 87–92 (1957).

Gluškov, V. M.

*5. On the central series of infinite groups. Mat. Sb. **31**, 491–496 (1952).

Golod, E. S.

*1. On nil-algebras and residually finite p-groups. Izv. Akad. Nauk. SSSR Ser. Mat. **28**, 273–276 (1964) = Amer. Math. Soc. Translations (2) **48**, 103–106 (1965).

Golod, E. S., Safarevič, I. R.

*1. On class field towers. Izv. Akad. Nauk. SSSR Ser. Mat. **28**, 261–272 (1964) = Amer. Math. Soc. Translations (2) **48**, 91–102 (1965).

Gorčakov, Yu. M.

*1. Embedding of locally normal groups in a direct product of finite groups. Dokl. Akad. Nauk. SSSR **137**, 26–28 (1961) = Soviet Math. Dokl. **2**, 514–516 (1961).

*2. Locally normal groups. Dokl. Akad. Nauk. SSSR **147**, 537–539 (1962) = Soviet Math. Dokl. **3**, 1654–1656 (1962).

*4. The existence of abelian subgroups of infinite rank in locally soluble groups. Dokl. Akad. Nauk. SSSR **156**, 17–20 (1964) = Soviet Math. Dokl. **5**, 591–594 (1964).

*5. On locally normal groups. Mat. Sb. **67**, 244–254 (1965).

Gorčinskiĭ, Yu. N.

*1. Groups with a finite number of conjugacy classes. Mat. Sb. **31**, 167–182 (1952).

*2. Periodic groups with a finite number of conjugacy classes. Mat. Sb. **31**, 209–216 (1952).

Gregorac, R. T.

1. A note on finitely generated groups. Proc. Amer. Math. Soc. **18**, 756–758 (1967).

de Groot, J.

1. Indecomposable abelian groups. Nederl. Akad. Wetensch. Proc. Ser. A **60**, 137–145 (1957).

Gruenberg, K. W.
3. The Engel elements of a soluble group. Illinois J. Math. **3**, 151–168 (1959).
4. The upper central series in soluble groups. Illinois J. Math. **5**, 436–466 (1961).
6. The Engel structure of linear groups. J. Algebra **3**, 291–303 (1966).

Grün, O.
1. Beiträge zur Gruppentheorie I. J. reine angew. Math. **174**, 1–14 (1935).

Gupta, C. K.
3. On certain soluble groups. Proc. Cambridge Philos. Soc. **66**, 1–4 (1969).

Guterman, M. M.
1. Normal systems. J. Algebra **4**, 317–320 (1966).

Haimo, F.
2. The *FC*-chain of a group. Canad. J. Math. **5**, 498–511 (1953).

Hall, M.
2. The theory of groups. New York: MacMillan 1959.

Hall, P.
1. A contribution to the theory of groups of prime-power order. Proc. London Math. Soc. (2) **36**, 29–95 (1934).
2. Verbal and marginal subgroups. J. reine angew. Math. **182**, 156–157 (1940).
4. Finiteness conditions for soluble groups. Proc. London Math. Soc. (3) **4**, 419–436 (1954).
5. Finite-by-nilpotent groups. Proc. Cambridge Philos. Soc. **52**, 611–616 (1956).
6. Nilpotent groups. Canad. Math. Congress Summer Sem. Univ. Alberta (1957): republished as "The Edmonton Notes on Nilpotent Groups", London: Queen Mary College Mathematics Notes (1969).
7. Some sufficient conditions for a group to be nilpotent. Illinois J. Math. **2**, 787–801 (1958).
8. Periodic *FC*-groups. J. London Math. Soc. **34**, 289–304 (1959).
9. Constructions for locally finite groups. J. London Math. Soc. **34**, 305–317 (1959).
10. On the finiteness of certain soluble groups. Proc. London Math. Soc. (3) **9**, 595–622 (1959).
11. The Frattini subgroups of finitely generated groups. Proc. London Math. Soc. (3) **11**, 327–352 (1961).
13. On non-strictly simple groups. Proc. Cambridge Philos. Soc. **59**, 531–553 (1963).

Hall, P., Hartley, B.
1. The stability group of a series of subgroups. Proc. London Math. Soc. (3) **16**, 1–39 (1966).

Hall, P., Kulatilaka, C. R.
1. A property of locally finite groups. J. London Math. Soc. **39**, 235–239 (1964).

Hallett, J. T., Hirsch, K. A.
1. Torsion-free groups having finite automorphism groups. I. J. Algebra **2**, 287–298 (1965).
2. Die Konstruktion von Gruppen mit vorgeschriebenen Automorphismengruppen. J. reine angew. Math. **239/240**, 32–46 (1969).

Hasse, H.
1. Zahlentheorie. Berlin: Akademie-Verlag 1949.

2. Vorlesungen über Zahlentheorie. 2. Aufl. Berlin/Göttingen/Heidelberg/New York: Springer 1964.

Head, T. J.
1. Note on the occurrence of direct factors in a group. Proc. Amer. Math. Soc. **15**, 193—195 (1964).

Heineken, H.
1. Eine Verallgemeinerung des Subnormalteilerbegriffs. Arch. Math. (Basel) **11**, 244—252 (1960).

Held, D.
4. On abelian subgroups of an infinite 2-group. Acta Sci. Math. (Szeged) **27**, 97—98 (1966).

Hering, C.
1. Gruppen mit nichttrivialer Trofimovzahl. Arch. Math. (Basel) **15**, 404—407 (1964).

Higman, G.
1. A finitely related group with an isomorphic proper factor group. J. London Math. Soc. **26**, 59—61 (1951).
2. A finitely generated infinite simple group. J. London Math. Soc. **26**, 61—64 (1951).
4. On infinite simple permutation groups. Publ. Math. Debrecen **3**, 221—226 (1954).
7. Subgroups of finitely presented groups. Proc. Roy. Soc. Ser. A **262**, 455—475 (1961).

Higman, G., Neumann, B. H., Neumann, H.
1. Embedding theorems for groups. J. London Math. Soc. **24**, 247—254 (1949).

Hirsch, K. A.
1. On infinite soluble groups I. Proc. London Math. Soc. (2) **44**, 53—60 (1938).
2. On infinite soluble groups II. Proc. London Math. Soc. (2) **44**, 336—344 (1938).
3. On infinite soluble groups III. Proc. London Math. Soc. (2) **49**, 184—194 (1946).
5. On infinite soluble groups IV. J. London Math. Soc. **27**, 81—85 (1952).
6. On infinite soluble groups V. J. London Math. Soc. **29**, 250—251 (1954).
7. Über lokal-nilpotente Gruppen. Math. Z. **63**, 290—294 (1955).

Hirsch, K. A., Zassenhaus, H.
1. Finite automorphism groups of torsion-free groups. J. London Math. Soc. **41**, 545—549 (1966).

Hulanicki, A.
1. Note on a paper of de Groot. Nederl. Akad. Wetensch. Proc. Ser. A **61**, 114 (1958).

Huppert, B.
1. Lineare auflösbare Gruppen. Math. Z. **67**, 479—518 (1957).

Jennings, S. A.
1. A note on chain conditions in nilpotent rings and groups. Bull. Amer. Math. Soc. **50**, 759—763 (1944).

Kalužnin (Kaloujnine), L. A.
1. Sur quelques propriétés des groupes d'automorphismes d'un groupe abstrait I. C. R. Acad. Sci. Paris **230**, 2067—2069 (1950).
3. Über gewisse Beziehungen zwischen einer Gruppe und ihrer Automorphismen. Berlin. Math. Tagung pp. 164—172 (1953).

*4. Locally normal groups of higher categories. Algebra i Mat. Logika: Studies in Algebra, Kiev, pp. 62–71 (1966).

Kaplansky, I.

1. Infinite abelian groups. Ann. Arbor: Univ. Michigan Press 1954.
2. An introduction to differential algebra. Actualités Sci. Industr. Paris: Hermann 1951.

Kargapolov, M. I.

*2. On the theory of semi-simple locally normal groups. Naučn. Dokl. Vysš. Fiz. Mat. Nauk. **6**, 3–7 (1958).

*11. On a problem of O. Yu. Schmidt. Sibirsk. Mat. Ž. **4**, 232–235 (1963).

Karrass, A., Solitar, D.

1. Some remarks on the infinite symmetric groups. Math. Z. **66**, 64–69 (1956).

Kegel, O. H.

3. Über den Normalisator von subnormalen und erreichbaren Untergruppen. Math. Ann. **163**, 248–258 (1966).
4. Über einfache, lokal endliche Gruppen. Math. Z. **95**, 169–195 (1967).
5. Noethersche 2-Gruppen sind endlich. Monats. Math. **71**, 424–426 (1967).
6. Zur Struktur lokal endlicher Zassenhausgruppen. Arch. Math. (Basel) **18**, 337–348 (1967).
7. Locally finite versus finite simple groups. Symposium, Theory of Finite Groups, Harvard Univ., pp. 247–249 (1969).
8. Lectures on locally finite groups. Math. Inst. Oxford (1969).

Kegel, O. H., Wehrfritz, B. A. F.

1. Strong finiteness conditions in locally finite groups. Math. Z. **117**, 309–324 (1970).

Kogalovskiĭ, S. R.

*1. Structure of characteristic universal classes. Sibirsk. Mat. Ž. **4**, 97–119 (1963).

*2. A theorem of Birkhoff. Uspehi Mat. Nauk. **20**, 206–207 (1965).

Kontorovič, P. G.

*7. Invariantly covered groups II. Mat. Sb. **28**, 79–88 (1951).

*8. Remarks on the hypercentre of a group. Ural. Gos. Univ. Mat. Zap. **23**, 27–29 (1960).

Kostrikin, A. I.

*1. On the connection between periodic groups and Lie rings. Izv. Akad. Nauk SSSR Ser. Mat. **21**, 289–310 (1957) = Amer. Math. Translations (2) **45**, 165–189 (1965).

*2. Lie rings satisfying the Engel condition. Izv. Akad. Nauk. SSSR Ser. Mat. **21**, 515–540 (1957) = Amer. Math. Soc. Translations (2) **45**, 191–220 (1965).

*4. The Burnside problem. Izv. Akad. Nauk. SSSR Ser. Mat. **23**, 3–34 (1959).

Kovács, L. G., Neumann, B. H.

2. On the existence of Baur soluble groups of arbitrary height. Acta Sci. Math. (Szeged) **26**, 143–144 (1965).

Kurata, Y.

1. A decomposition of normal subgroups in a group. Osaka J. Math. **1**, 201–229 (1964).

Kuroš, A. G.

2. Eine Verallgemeinerung des Jordan-Hölderschen Satz. Math. Ann. **111**, 13–18 (1935).

*8. Radicals of rings and algebras. Mat. Sb. **33**, 13–26 (1953).

9. The theory of groups, 2nd ed. (2 vols.), New York: Chelsea 1960.

10. Radicaux en theorie des groupes. Bull. Soc. Math. Belg. **14**, 307–310 (1962).
*11. Radicals in the theory of groups. Sibirsk. Mat. Ž. 3, 912–931 (1962).
*12. Lectures on general algebra. Moscow: Izdat. Fiz-Mat. 1962 = General Algebra. New York: Chelsea 1963.
*13. The theory of groups. 3rd augmented ed. (3 vols.), Moscow: Izdat. Nauka 1967.

Kuroš, A. G., Černikov, S. N.
*1. Soluble and nilpotent groups. Uspehi Mat. Nauk. **2**, 18–59 (1947) = Amer. Math. Soc. Translations **80** (1953).

Lam, T.-Y.
1. A commutator formula for a pair of subgroups and a theorem of Blackburn. Canad. Math. Bull. **12**, 217–219 (1969).

Lennox, J. C., Roseblade, J. E.
1. Centrality in finitely generated soluble groups. J. Algebra **16**, 399–435 (1970).

Levi, F. W.
1. Über die Untergruppen freier Gruppen. Math. Z. **37**, 90–97 (1933).

Levin, F.
2. One variable equations over groups. Arch. Math. (Basel) **15**, 179–188 (1964).
5. Factor groups of the unimodular group. J. London Math. Soc. **43**, 195–203 (1968).

Livčak, Ya. B.
*2. On the theory of generalized soluble groups. Sibirsk. Mat. Ž. **1**, 617–622 (1960).

Lyndon, R. C., Ullman, J. L.
1. Pairs of real 2-by-2 matrices that generate free products. Michigan Math. J. **15**, 161–166 (1968).

McCarthy, D.
1. Infinite groups whose proper quotient groups are finite I. Comm. Pure Appl. Math. **21**, 545–562 (1968).
2. Infinite groups whose proper quotient groups are finite II. Comm. Pure Appl. Math. **23**, 767–789 (1970).

McCool, J.
2. Unsolvable problems in groups with solvable word problem. Canad. J. Math. **22**, 836–838 (1970).

Macdonald, I. D.
1. A class of *FC*-groups. J. London Math. Soc. **34**, 73–80 (1959).
2. Some explicit bounds in groups with finite derived groups. Proc. London Math. Soc. (3) **11**, 23–56 (1961).

McDougall, D.
2. Soluble groups with the minimum condition for normal subgroups. Math. Z. **118**, 157–167 (1970).

MacHenry, T.
1. The tensor product and the 2nd nilpotent product of groups. Math. Z. **73**, 134–145 (1960).

McLain, D. H.
3. On locally nilpotent groups. Proc. Cambridge Philos. Soc. **52**, 5–11 (1956).
4. Remarks on the upper central series of a group. Proc. Glasgow Math. Assoc. **3**, 38–44 (1956).

5. Finiteness conditions in locally soluble groups. J. London Math. Soc. **34**, 101–107 (1959).
6. Local theorems in universal algebras. J. London Math. **34**, 177–184 (1959).

Magnus, W., Karrass, A., Solitar, D.
1. Combinatorial group theory. New York: Wiley 1966.

Mal'cev, A. I.
*1. On the faithful representation of infinite groups by matrices. Mat. Sb. **8**, 405–422 (1940) = Amer. Math. Soc. Translations (2) **45**, 1–18 (1965).
*2. On a general method for obtaining local theorems in group theory. Ivanov. Gos. Ped. Inst. Učen. Zap. **1**, 3–9 (1941).
*3. On groups of finite rank. Mat. Sb. **22**, 351–352 (1948).
*4. Nilpotent torsion-free groups. Izv. Akad. Nauk. SSSR Ser. Mat. **13**, 201–212 (1949).
*7. On certain classes of infinite soluble groups. Mat. Sb. **28**, 567 –588 (1951) = Amer. Math. Soc. Translations (2) **2**, 1–21 (1956).

Medvedeva, R. P.
*1. A generalization of finite groups with the transitive property for normal subgroups. Sibirsk. Mat. Ž. **6**, 1068–1073 (1965).

Meldrum, J. D. P.
1. On central series of a group. J. Algebra **6**, 281–284 (1967).

Mennicke, J. L.
1. Finite factor groups of the unimodular group. Ann. of Math. (2) **81**, 31–37 (1965).

Menogazzo, F.
1. Gruppi nei quali la relazione di quasi-normalità è transitiva. Rend. Sem. Mat. Univ. Padova **40**, 347–361 (1968).
2. Gruppi nei quali la relazione di quasi-normalità è transitiva II. Rend. Sem Mat. Univ. Padova **42**, 389–399 (1969).

Merzljakov, Yu. I.
*2. Locally soluble groups of finite rank. Algebra i Logika **3**, 5–16 (1964).
*4. Verbal and marginal subgroups of linear groups. Dokl. Akad. Nauk. SSSR **177**, 1008–1011 (1967) = Soviet Math. Dokl. **8**, 1538–1541 (1967).

Miller, G. A., Moreno, H. C.
1. Non-abelian groups in which every subgroup is abelian. Trans. Amer. Math. Soc. **4**, 398–404 (1903).

Neumann, B. H.
1. Identical relations in groups I. Math. Ann. **114**, 506–525 (1937).
2. Some remarks on infinite groups. J. London Math. Soc. **12**, 120–127 (1937).
6. A two-generator group isomorphic to a proper factor group. J. London Math. Soc. **25**, 247–248 (1950).
7. Groups with finite classes of conjugate elements. Proc. London Math. Soc. (3) **1**, 178–187 (1951).
9. Groups covered by permutable subsets. J. London Math. Soc. **29**, 236–248 (1954).
11. Groups covered by finitely many cosets. Publ. Math. Debrecen **3**, 227–242 (1954).
12. Groups with finite classes of conjugate subgroups. Math. Z. **63**, 76–96 (1955).

Neumann, B. H., Neumann, H.
2. Embedding theorems for groups. J. London Math. Soc. **34**, 465–479 (1959).

Neumann, H.
1. Varieties of groups. Berlin/Heidelberg/New York: Springer 1967.

Neumann, P. M.
2. An improved bound for *BFC* *p*-groups. J. Austral. Math. Soc. **11**, 19–27 (1970).

Newell, M. L.
4. Finiteness conditions in generalized soluble groups. J. London Math. Soc. (2) **2**, 593–596 (1970).

Newman, M.
1. Pairs of matrices generating discrete free groups and free products. Michigan Math. J. **15**, 155–160 (1968).

Newman, M. F.
4. Another non-Hopf group. J. London Math. Soc. **41**, 292 (1966).

Newman, M. F., Wiegold, J.
1. Groups with many nilpotent subgroups. Arch. Math. (Basel) **15**, 241–250 (1964).

Nishigôri, N.
2. On *FC*-solvable groups. J. Sci. Hiroshima Univ. Ser. A–I **25**, 367–368 (1961).

Nisnevič, V. L.
*1. On groups which are isomorphically representable by matrices over a commutative field. Mat. Sb. **8**, 395–403 (1940).

Northcott, D. G.
1. Ideal theory. London: Cambridge Univ. Press 1953.

Novikov, P. S., Adjan, S. I.
*1. Infinite periodic groups. Izv. Akad. Nauk. SSSR Ser. Mat. **32**, 212–244, 251–524, 709–731 (1968) = Math. USSR-Izv. **2**, 209–236, 241–479, 665–685 (1968).
*2. Commutative subgroups and the conjugacy problem in free periodic groups of odd order. Izv. Akad. Nauk. SSSR Ser. Mat. **32**, 1176–1190 (1968) = Math. USSR-Izv. **2**, 1131–1144 (1968).

Peng, T. A.
2. Finite groups with pro-normal subgroups. Proc. Amer. Math. Soc. **20**, 232–234 (1969).

Phillips, R. E., Combrink, C. R.
1. A note on subsolvable groups. Math. Z. **92**, 349–352 (1966).

Phillips, R. E., Robinson, D. J. S., Roseblade, J. E.
1. Maximal subgroups and chief factors of certain generalized soluble groups. Pacific J. Math. **37**, 475–480 (1971).

Pic, G.
2. Une propriété des groupes *FC*-nilpotents. Com. Acad. RPR **12**, 969–972 (1962).

Platonov, V. P.
*6. On a problem of Mal'cev. Mat. Sb. **79**, 621–624 (1969) = Math. USSR-Sb. **8**, 599–602 (1969).

Plotkin, B. I.
*6. On some criteria of locally nilpotent groups. Uspehi Mat. Nauk. **9**, 181–186 (1954) = Amer. Math. Soc. Transl. (2) **17**, 1–7 (1961).

*7. Radical groups. Mat. Sb. **37**, 507—526 (1955) = Amer. Math. Soc. Translations (2) **17**, 9—28 (1961).

*10. On groups with finiteness conditions for abelian subgroups. Dokl. Akad. Nauk. SSSR **107**, 648—651 (1956).

*11. Radical and semi-simple groups. Trudy Moskov. Mat. Obšč. **6**, 299—336 (1957).

*12. Generalized soluble and nilpotent groups. Uspehi Mat. Nauk. **13**, 89—172 (1958) = Amer. Math. Soc. Translations (2) **17**, 29—115 (1961).

*16. Radical groups whose radical has an ascending central series. Ural. Gos. Univ. Mat. Zap. **23**, 40—43 (1960).

*18. Some properties of automorphisms of nilpotent groups. Dokl. Akad. Nauk. SSSR **137**, 1303—1306 (1961) = Soviet Math. Dokl. **2** (1961), 471—474 (1961).

*25. Infinite-dimensional linear groups. Dokl. Akad. Nauk. SSSR **153**, 42—45 (1963) = Soviet Math. Dokl. **4**, 1617—1620 (1963).

Polovickiĭ, Ya. D.

*1. Layer-extremal groups. Dokl. Akad. Nauk. SSSR **134**, 533—535 (1960) = Soviet Math. Dokl. **1**, 1112—1113 (1960).

*2. On locally extremal groups and groups with the π-minimal condition. Dokl. Akad. Nauk. SSSR **138**, 1022—1024 (1961) = Soviet Math. Dokl. **2**, 780—782 (1961).

*3. Layer-extremal groups. Mat. Sb. **56**, 95—106 (1962).

*4. Locally extremal and layer-extremal groups. Mat. Sb. **58**, 685—694 (1962).

*6. A condition for abelian groups. Perm. Gos. Univ. Učen. Zap. **22**, 41—42 (1962).

*7. Groups with extremal classes of conjugate elements. Sibirsk. Mat. Ž. **5**, 891—895 (1964).

Remak, R.

1. Über minimale invariante Untergruppen in der Theorie der endlichen Gruppen. J. reine angew. Math. **162**, 1—16 (1930).

Rhemtulla, A. H.

3. A property of groups with no central factors. Canad. Math. Bull. **12**, 467—470 (1969).

Rips, I. A.

*1. Two propositions about Baer groups. Dokl. Akad. Nauk. SSSR **186**, 264—267 (1969) = Soviet Math. Dokl. **10**, 589—592 (1969).

Robinson, D. J. S.

1. Groups in which normality is a transitive relation. Proc. Cambridge Philos. Soc. **60**, 21—38 (1964).

2. Joins of subnormal subgroups. Illinois J. Math. **9**, 144—168 (1965).

3. On finitely generated soluble groups. Proc. London Math. Soc. (3) **15**, 508—516 (1965).

4. On the theory of subnormal subgroups. Math. Z. **89**, 30—51 (1965).

5. Wreath products and indices of subnormality. Proc. London Math. Soc. (3) **17**, 257—270 (1967).

7. Residual properties of some classes of infinite soluble groups. Proc. London Math. Soc. (3) **18**, 495—520 (1968).

8. A note on finite groups in which normality is transitive. Proc. Amer. Math. Soc. **19**, 933—937 (1968).

9. Finiteness conditions for subnormal and ascendant abelian subgroups. J. Algebra **10**, 333—359 (1968).

10. A property of the lower central series of a group. Math. Z. **107**, 225–231 (1968).
11. Infinite soluble and nilpotent groups. London: Queen Mary College Mathematics Notes (1968).
13. Groups which are minimal with respect to normality being intransitive. Pacific J. Math. **31**, 777–785 (1969).
14. On the theory of groups with extremal layers. J. Algebra **14**, 182–193 (1970).

Roseblade, J. E.
3. On certain subnormal coalition classes. J. Algebra **1**, 132–138 (1964).
4. The permutability of orthogonal subnormal subgroups. Math. Z. **90**, 365–372 (1965).
5. A note on subnormal coalition classes. Math. Z. **90**, 373–375 (1965).
7. A note on disjoint subnormal subgroups. Bull. London Math. Soc. **1**, 65–69 (1969).

Roseblade, J. E., Stonehewer, S. E.
1. Subjunctive and locally coalescent classes of groups. J. Algebra **8**, 423–435 (1968).

Rosenlicht, M.
1. On a result of Baer. Proc. Amer. Math. Soc. **13**, 99–101 (1962).

Šain, B. M.
*1. On the Birkhoff-Kogalovskiĭ theorem. Uspehi Mat. Nauk. **20**, 173–174 (1965).

Saşiada, E.
1. Construction of directly indecomposable abelian groups of a power higher than that of the continuum. Bull. Acad. Polon. Sci. Sér. Sci. Math. Astronom. Phys. **5**, 701–703 (1957), **7**. 23–26 (1959).

Schenkman, E.
5. The similarity between the properties of ideals in commutative rings and the properties of normal subgroups of groups. Proc. Amer. Math. Soc. **9**, 375–381 (1958).
7. Group theory. Princeton: Van Nostrand 1965.

Schiefelbusch, L.
1. The Trofimov number of some infinite groups with finiteness conditions. Arch. Math. (Basel) **18**, 122–127 (1967).

Schlette, A.
1. Artinia, nalmost abelian groups and their groups of automorphisms. Pacific J. Math. **29**, 403–425 (1969).

Schmidt, O. J. (Šmidt, O. Yu.)
*2. On groups every proper subgroup of which is special. Mat. Sb. (O. S.) **31**, 366–372 (1924).
*5. Infinite soluble groups. Mat. Sb. **17**, 145–162 (1945).
*6. Selected works. Moscow: Matematika 1959.
7. Die lokale Endlichkeit einer Klasse unendlicher periodischer Gruppen. Math. Forschungsberichte XX. Berlin: VEB Deutscher Verlag der Wissenschaften 1963, pp. 79–81.

Schur, I.
1. Neuer Beweis eines Satzes über endliche Gruppen. Sitzber. Akad. Wiss. Berlin, pp. 1013–1019 (1902).
2. Über die Darstellungen der endlichen Gruppen durch gebrochene lineare Substitutionen. J. reine angew. Math. **127**, 20–50 (1904).

4. Über Gruppen periodischer linearer Substitutionen. Sitzber. Preuss. Akad. Wiss. pp. **619**–627 (1911).

Scott, W. R.

3. On a result of B. H. Neumann. Math. Z. **66**, 240 (1956).
5. Group theory. Englewood Cliffs: Prentice Hall 1964.

Ščukin, K. K.

*4. An RI^*-soluble radical of groups. Dokl. Akad. Nauk. SSSR **132**, 541–544 (1960) = Soviet Math. Dokl. **1**, 615–618 (1960).
*6. On the theory of radicals in groups. Dokl. Akad. Nauk. SSSR **142**, 1047–1049 (1962) = Soviet Math. Dokl. **3**, 260–263 (1962).
*8. On the theory of radicals in groups. Sibirsk. Mat. Ž. **3**, 932–942 (1962).

Shepperd, J. A. H., Wiegold, J.

1. Transitive permutation groups and groups with finite derived groups. Math. Z. **81**, 279–285 (1963).

Simon, H.

1. Noethersche Gruppen mit endlicher Hyperzentrumsfaktorgruppe. Illinois J. Math. **8**, 231–240 (1964).
3. Eine Verallgemeinerung eines Satzes von Mal'cev und Baer. Arch. Math. (Basel) **17**, 289–291 (1966).

Skolem, T.

1. On the existence of a multiplicative basis for an arbitrary algebraic field. Norske Vid. Selsk. Forh. (Trondheim) **20**, 4–7 (1947).

Šmel'kin, A. L.

*2. A property of semi-simple classes of groups. Sibirsk. Mat. Ž. **3**, 950–951 (1962).
*5. On soluble products of groups. Sibirsk. Mat. Ž. **6**, 212–220 (1965).

Smirnov, D. M.

*4. On groups of automorphisms of soluble groups. Mat. Sb. **32**, 365–384 (1953).

Specht, W.

2. Beiträge zur Gruppentheorie I. Lokalendliche Gruppen. Math. Nachr. **18**, 39–56 (1958).

Stewart, A. G. R.

1. On the class of certain nilpotent groups. Proc. Roy. Soc. Ser. A **292**, 374–379 (1966).

Stewart, I. N.

1. The minimal condition for subideals of Lie algebras. Math. Z. **111**, 301–310 (1969).

Stroud, P. W.

1. On a property of verbal and marginal subgroups. Proc. Cambridge Philos. Soc. **61**, 41–48 (1965).

Strunkov, S. P.

*3. Normalizers and abelian subgroups of certain classes of groups. Izv. Akad. Nauk. SSSR Ser. Mat. **31**, 657–670 (1967) = Math. USSR-Izv. **1**, 639–650 (1967).

Sulinski, A., Anderson, R., Divinsky, N.

1. Lower radical properties for associative and alternative rings. J. London Math. Soc. **41**, 417–424 (1966).

Šunkov, V. P.

*3. Abstract characterization of a simple projective group of type $PGL(2, K)$ over a field K of characteristic $r \neq 0$ or 2. Dokl. Akad. Nauk. SSSR **163**, 837–840 (1965) = Soviet Math. Dokl. **6**, 1043–1047 (1965).

*4. A contribution to the theory of locally finite groups. Dokl. Akad. Nauk. SSSR **168**, 1272–1274 (1966) = Soviet Math. Dokl. **7**, 841–843 (1966).
*5. On the theory of periodic groups. Dokl. Akad. Nauk. **175**, 1236–1237 (1967) = Soviet Math. Dokl. **8**, 1011–1012 (1967).
*6. A locally finite group with extremal Sylow p-subgroups for someprime p. Sibirsk. Mat. Ž. **8**, 213–229 (1967) = Siberian Math. J. **8**, 161–171 (1967).
*8. Minimality problem for subgroups in locally finite groups. Dokl. Akad. Nauk. SSSR **181**, 294–295 (1968) = Soviet Math. Dokl. **9**, 840–842 (1968).
*10. On the minimality problem for locally finite groups. Algebra i Logika **9**, 220–248 (1970) = Algebra and Logic **9** 137–151 (1970).

Suprunenko, D. A.
*1. Soluble and nilpotent linear groups. Minsk (1958) = Amer. Math. Soc. Translations of Mathematical Monographs **9**, (1963).

Szász, F.
1. Groups in which every non-trivial power is cyclic. Magyar Tud. Akad. Fiz. Oszt. Közl. **5**, 491–492 (1955).
2. Über Gruppen, deren sämtliche nicht-triviale Potenzen zyklische Untergruppen sind. Acta Math. Sci. (Szeged) **17**, 83–84 (1955).
3. A characterization of the cyclic troups. Rev. Roumaine Math. Pures Appl. **1**, 13–16 (1956).
4. On cyclic groups. Fund. Math. **43**, 238–240 (1956).
6. Bemerkung zu meiner Arbeit "Über Gruppen deren sämtliche nicht-triviale Potenzen zyklische Untergruppen sind". Acta Sci. Math. (Szeged) **23**, 64–66 (1962).

Szép, J., Itô, N.
1. Über die Faktorisation von Gruppen. Acta Sci. Math. (Szeged) **16**, 229–231 (1955).

Tokarenko, A. I.
*1. On linear groups over rings. Sibirsk. Mat. Ž. **9**, 951–959 (1968) = Siberian Math. J. **9**, 708–713 (1968).

Trofimov, P. I.
*1. A study of the influence of the greatest common divisor of the orders of the classes of conjugate non-normal Sylow subgroups of a finite group and its properties. Sibirsk. Mat. Ž. **4**, 236–239 (1963).

Turner-Smith, R. F.
1. Marginal subgroup properties for outer commutator words. Proc. London Math. Soc. (3) **14**, 321–341 (1964).
2. Finiteness conditions for verbal subgroups. J. London Math. Soc. **41**, 166–176 (1966).

Valiev, M. K.
*1. A theorem of G. Higman. Algebra i Logika **7**, 9–22 (1968) = Algebra and Logic **7**, 135–143 (1968).

de Vries, H., de Miranda, A. B.
1. Groups with a small number of automorphisms. Math. Z. **68**, 450–464 (1958).

Wehrfritz, B. A. F.
8. Sylow subgroups of locally finite groups with Min-p. J. London Math. Soc. (2) **1**, 421–427 (1969).

Weidig, I.
1. Gruppen mit abgeschwächter Normalteilertransitivität. Rend. Sem. Mat. Univ. Padova **36**, 185–215 (1966).

Wiegold, J.
1. Groups with boundedly finite classes of conjugate elements. Proc. Roy. Soc. London Ser. A **238**, 389–401 (1957).
2. Nilpotent products of groups with amalgamations. Publ. Math. Debrecen **6**, 131–168 (1959).
3. On direct factors in groups. J. London Math. Soc. **35**, 310–320 (1960).
6. Multiplicators and groups with finite central factor-groups. Math. Z. **89**, 345–347 (1965).
8. Commutator subgroups of finite p-groups. J. Austral. Math. Soc. **10**, 480–484 (1969).

Wielandt, H.
2. Eine Verallgemeinerung der invarianten Untergruppen. Math. Z. **45**, 209–244 (1939).
3. Vertauschbare nachinvariante Untergruppen. Abh. Math. Sem. Univ. Hamburg **21**, 55–62 (1957).
4. Über den Normalisator der subnormalen Untergruppen. Math. Z. **69**, 463–465 (1958).

Wilson, J. S.
1. Some properties of groups inherited by normal subgroups of finite index. Math. Z. **114**, 19–21 (1970).
3. Groups with every proper quotient finite. Proc. Cambridge Philos. Soc. **69**, 373–392 (1971).

Witt, E.
2. Über die Kommutatorgruppe kompakter Gruppen. Rend. Mat. e Appl. (5) **14**, 125–129 (1954).

Zacher, G.
1. Caratterizzazione dei t-gruppi finiti risolubili. Ricerche Mat. **1**, 287–294 (1952).

Zaičev, D. I.
*4. Groups satisfying a weak minimum condition. Dokl. Akad. Nauk. SSSR, **178**, 780–782 (1968) = Soviet Math. Dokl. **9**, 194–197 (1968).
*6. Groups which satisfy a weak minimality condition. Ukrain. Mat. Ž. **20**, 472–482 (1968) = Ukrainian Math. J. **20** (1968).
*8. On groups which satisfy a weak minimality condition. Mat. Sb. **78**, 323–331 (1969) = Math. USSR-Izv. **7**, 315–322 (1969).

Zappa, G.
1. Sui gruppi di Hirsch supersolubili I. Rend. Sem. Mat. Univ. Padova **12**, 1–11 (1941).
2. Sui gruppi di Hirsch supersolubili II. Rend. Sem. Mat. Univ. Padova **12**, 62–80 (1941).

Zassenhaus, H.
1. Beweis eines Satzes über diskrete Gruppen. Abh. Math. Sem. Univ. Hamburg **12**, 289–312 (1938).
3. The theory of groups. 2nd ed. New York: Chelsea 1958.

Zel'manzon, M. E.
*1. Groups in which all subgroups are cyclic. Uspehi Mat. Nauk **16**, 109–113 (1961).

Author Index

Abramovskiĭ, I. N. 174
Adjan, S. I. 35, 143
Ado, I. D. 90, 154
Alperin, J. L. 98, 108
Amberg, B. 98, 99, 180
Amitsur, S. 18
Anderson, R. 22
Auslander, L. 83

Bačurin, G. F. 103
Baer, R. 3, 12, 14, 18, 22, 34, 41, 47, 50, 55, 57, 58, 59, 62, 63, 69, 72, 73, 82, 83, 84, 85, 86, 89, 91, 92, 101, 102, 103, 104, 105, 107, 108, 111, 113, 121, 122, 131, 132, 134, 138, 139, 144, 146, 148, 149, 151, 152, 153, 155, 158, 166, 167, 171, 176, 177, 180
Balcerzyk, S. 27
Bass, H. 173
Baumslag, G. 33, 40, 83
Bender, H. 98
Berlinkov, M. L. 139
Berman, S. D. 17
Best, E. 174
Betten, A. 57
Birkhoff, G. 7, 18
Blackburn, N. 55
Bowers, J. F. 67
Brauer, R. 98
Brenner, J. L. 40, 173
Bride, I. M. 127
Burnside, W. 35, 95, 96, 97, 129

Calenko, M. S. 143, 144
Camm, R. 143
Čan Van Hao 23, 25
Čarin, V. S. 36, 152, 153, 154, 156, 157, 167
Černikov, S. N. 18, 50, 53, 68, 74, 84, 85, 86, 99, 100, 121, 122, 125, 134, 135, 136, 138, 139, 140, 151, 152, 156
Chehata, C. G. 17, 144
Clapham, C. R. J. 32
Clifford, A. H. 75, 157
Clowes, J. S. 144
Combrink, C. R. 22
Curtis, C. 39, 85, 129, 164

Dark, R. S. 30, 166, 185
Dedekind, R. 10, 39
Dickson, S. E. 27
Dietzmann, A. P. 45, 122, 134, 149
Dirichlet, P. G. L. 81, 110, 140
Divinsky, N. 22
Dixmier, S. 52, 55
Dixon, J. D. 78
Dlab, V. 134
Duguid, A. M. 129, 131, 132, 148, 152

Erdös, J. 121, 123
Eremin, I. I. 127, 139

Fedorov, Yu. G. 123
Feit, W. 95, 96, 100
Fitting, H. 49, 57, 133
Frobenius, G. 95
Fuchs, L. 34, 68, 79, 108, 135, 141, 152, 157
Gaschütz, W. 174
Gluškov, V. M. 131
Golod, E. S. 35
Gorčakov, Yu. M. 100, 124, 125
Gorčinskiĭ, Yu. N. 129
Gorenstein, D. 98
Gregorac, R. J. 31
de Groot, J. 108
Gruenberg, K. W. 12, 58, 59, 79, 80
Grün, O. 48
Gupta, C. K. 166
Guterman, M. M. 17

Haimo, F. 129, 131, 133
Hall, M. 101
Hall, P. 3, 9, 10, 13, 17, 25, 26, 30, 32, 40, 44, 51, 56, 57, 58, 72, 81, 92, 93, 95, 97, 98, 100, 101, 102, 111, 112, 113, 114, 115, 116, 117, 119, 121, 123, 124, 125, 140, 144, 146, 152, 160, 161, 163, 164
Hallett, J. T. 111
Hartley, B. 25
Hasse, H. 81, 110, 140
Head, T. J. 178
Heineken, H. 12
Held, D. 93
Hering, C. 127
Higman, G. 30, 32, 40, 129, 143, 144, 145
Hirsch, K. A. 57, 58, 62, 63, 65, 66, 67, 111, 144
Hölder, O. 16, 17, 66, 184
Hopf, H. 40

Hulanicki, A. 108
Huppert, B. 78

Itô, N. 45
Iwasawa, K. 41

Jennings, S. A. 152, 160
Jordan, C. 16, 17, 66, 184

Kalužnin, L. A. 37, 44
Kaplansky, I. 102, 109, 110
Kargapolov, M. I. 36, 92, 93, 95, 97, 98, 100, 125, 174
Karrass, A. 33, 35, 144
Kegel, O. H. 40, 70, 93, 98, 185
Kogalovskiĭ, S. R. 7
Kontorovič, P. G. 37, 53
Kostrikin, A. I. 35, 143
Kovács, L. G. 13, 36
Kulatilaka, C. R. 92, 93, 95, 97, 98, 100
Kurata, Y. 169, 170
Kuroš, A. G. 5, 10, 11, 17, 18, 20, 22, 23, 25, 26, 27, 31, 35, 38, 51, 68, 81, 171, 184

Lam, T.-Y. 55
Lazard, M. 173
Lennox, J. C. 166
Levi, F. W. 41
Levin, F. 30
Livčak, Ya. B. 25
Loewy, A. 151
Lyndon, R. C. 40
Lyubimov, V. V. 17

Macdonald, I. D. 119, 125, 127
MacHenry, T. 125
Magnus, W. 33, 35
Mal'cev, A. I. 34, 39, 41, 50, 53, 75, 79, 80, 82, 86, 95, 97, 100, 154, 157
McCarthy, D. 123
McCool, J. 32
McDougall, D. 153
McLain, D. H. 5, 52, 53, 129, 130, 131, 132, 133, 148, 152, 154, 155, 156, 160, 166, 167, 168
Medvedeva, R. P. 174
Meldrum, J. D. P. 48
Mennicke, J. L. 173
Menogazzo, F. 174
Merzljakov, Yu. I. 100, 112, 121
Miller, G. A. 98
de Miranda, A. B. 111
Moreno, H. C. 98

Neumann, B. H. 8, 13, 30, 31, 32, 36, 40, 102, 105, 107, 121, 122, 123, 126, 127, 129, 145
Neumann, H. 8, 30, 32, 40, 129, 145
Neumann, P. M. 127
Newell, M. L. 158
Newman, M. 40
Newman, M. F. 40, 97
Nishigôri, N. 132, 133
Nisnevič, V. L. 40
Northcott, D. G. 170
Novikov, P. S. 35, 143

Peng, T. A. 174
Phillips, R. E. 22, 167
Pic, G. 133
Platonov, V. P. 34
Plotkin, B. I. 3, 12, 17, 18, 40, 57, 58, 59, 62, 63, 74, 78, 86, 156, 180
Polovickiĭ, Ya. D. 74, 84, 115, 122, 127, 129, 133, 134, 135, 139
Prüfer, H. 33, 34

Reidemeister, K. 31, 102
Reiner, I. 39, 85, 129, 164
Remak, R. 149, 150, 178
Rhemtulla, A. H. 166, 168
Rips, I. A. 62, 64
Robinson, D. J. S. 15, 54, 55, 57, 60, 135, 136, 138, 139, 157, 158, 167, 171, 173, 174, 175, 176, 177, 181, 182, 185, 186
Roseblade, J. E. 60, 136, 153, 166, 167, 171, 175, 177, 182, 185, 186
Rosenlicht, M. 104

Šafarevič, I. R. 35
Šain, B. M. 7
Saşiada, E. 108
Schenkman, E. 125, 169, 170
Schiefelbusch, L. 127
Schlette, A. 84, 111, 115
Schmidt, O, J. 35, 71 86, 92, 97, 98
Schreier, O. 31, 66, 102, 184
Schur, I. 39, 81, 85, 101, 102, 103
Scott, W. R. 90, 96, 97, 109, 127, 150
Ščukin, K. K. 14, 21, 22, 47
Serre, J.-P. 173
Shepperd, J. A. H. 127
Simon, H. 74, 92
Skolem, T. 157
Šmel'kin, A. L. 25, 165
Smirnov, D. M. 82, 85
Solitar, D. 33, 35, 40, 144
Specht, W. 36
Stewart, A. G. R. 57
Stewart, I. N. 186
Stonehewer, S. E. 60, 171, 175
Stroud, P. W. 112, 113, 115
Strunkov, S. P. 93, 97
Sulinski, A. 22
Šunkov, V. P. 93, 95, 96, 97, 98, 100
Suprunenko, D. A. 74, 75
Szász, F. 134
Szép, J. 45

Tarski, A. 97
Taussky, O. 174
Thompson, J. G. 95, 96, 100
Tokarenko, A. I. 39
Trofimov, P. I. 127

series, composition, weak 66
—, descending 10—11
—, normal 12
— of successive normal closures 173
—, principal 17
—, subnormal 12
—, upper socle 151
—, χ- 13
simple, absolutely 16
—, characteristically 18
— groups, direct products of 178ff.
— —, embedding in 144ff.
— —, infinite 143ff.
—, strictly 17
— subgroup, subnormal 177ff.
socle, 149, 178
—, abelian 150
—, non-abelian 150
—, subnormal 180ff.
soluble group 45
— — with Max-n 160
— — with Min-n 152—153
soluble length 45
solvable group 45
subgroup theoretical class 9
— — property 9
subnormal 12
— index 173
— —, groups with bounded 173ff.
— intersection property 175
— join property 175
subsoluble group 22, 176
Šunkov, theorem of 93
supersoluble 66
—, locally 73, 156, 169

tensor product 54
thin 138
Three Subgroup Lemma 44
torsion group 34
transfer 101
transversal, right 30
triangulable 74
Trofimov number 127

unary operation 6
unimodular group 81
unitriangulable 74
universal class 7

variety 7
— generated by a class of groups 8
verbal mapping 8
— subgroup 8, 111ff.

Wielandt subgroup 177ff.
—, theorem of 59, 175
word, commutator 112
—, —, outer 112
—, concise 119
—, derived 119
—, lower central 113
—, verbose 119

$\mathfrak{X}$-group 1

Z-group 49
ZA-group 14, 49
ZD-group 49
Zassenhaus' Lemma 17
— theorem on soluble linear groups 78

Subject Index

A-closed 3
artinian group 38
ascendant 12
ascending degree α, groups of 22
automorphism groups, groups with finite 108 ff.
— —, irreducible 80, 153
— — of Černikov groups 83 ff.
— — of polyclic-by-finite groups 82
— — of soluble groups 74 ff.
— —, rationally irreducible 80, 159

BFC-group 126
Baer, criterion of 72
— group 61
— radical 61
—, theorem of 57, 59, 73, 103—105, 144
Burnside group 35, 143
— problem 35, 68
— —, restricted 35

CC-group 127—129
CL-group 134 ff.
category, groups with a 36, 37, 64
—, — — — abelian 64
—, — — — finite 36
central-by-finite group 101, 102, 105, 139
central product 117
— series 48
— —, lower χ- 29
— —, upper χ- 28
centre, 28
—, *FC*- 121
—, χ- 27
Černikov conjugacy classes 127—129
— group 67, 164
— —, minimal non- 98

chain condition 37
— —, ascending 37
— —, descending 37
class of groups 1
Clifford's Theorem 75
closure operation 3
— — generated by a set of operations 5
— —, identity 4
— —, relative 7
— —, universal 4
coalition class, ascendant 59, 139
— —, subnormal 59, 139
co-hopfian 41
commutator 42
— subgroup 43
conjugacy classes, finite 121 ff.
— —, finitely many 129
core 60
covering 105
—, finite 105—107, 123

defect 173
derived length 45
— series 45
descendant 12
descending degree α, groups of 25
diagonable 74
Dietzmann's Lemma 45, 122, 134 149
direct product of Černikov groups 135 ff.
— — of finite groups 123 ff., 139
— —, prime thin 135
disjoint classes 7

exponent 34
extension class 2
extremal group 67

FC-centre 121
— element 121
— group 121
— —, periodic 123 ff.
— hypercentral group 129
— hypercentre 129, 149, 159
— nilpotent group 129
— soluble group 129
FL-group 134 ff.
factor, chief 17, 154
—, composition 16, 180
— of a series 10
Fedorov, theorem of 123
Feit-Thompson Theorem 95, 96
finite-by-nilpotent group 117
finitely generated group 30
finitely presented group 31
finiteness condition 29
— — on conjugates and commutators 101 ff.
— — on subnormal subgroups 170 ff.
— criteria 45
— — for commutator groups 103—105
Fitting subgroup 18
Frobenius group 95, 96

group theoretical class 1
— — function 18
— — functional 37
— — property 1
Gruenberg group 61
— radical 61
—, theorem of 59
Grün's Lemma 48

Hall-Kulatilaka-Kargapolov theorem 92
— — — —, generalization of 97
Hall's problems 112 ff.
— type, theorems of 57
Hilbert's Basis Theorem 161
Hirsch-Plotkin-Baer Theorem 57
Hirsch-Plotkin radical 58
hopfian group 40
hyperabelian 46
— central 14, 49
— cyclic 73
— finite 36
— χ 14
hypercentre 28, 130, 149
—, *FC*- 129, 149, 159
hypercentre, χ- 28
hypoabelian 25
— central 15, 49
— χ 15
hypocentre 29
—, χ- 29

involution 70
isomorphism classes of subgroups, groups with finite 141 ff.
Iwasawa's theorem 41

Jordan-Hölder Theorem 16, 17, 66, 184
just-infinite group 123, 171 ff.

Kuroš correspondence 25, 26

layer 133
—, abelian 133
—, Černikov 134
—, cyclic 134
—, finite 134
—, polycyclic-by-finite 139
linear groups 39—40
— —, soluble 74 ff., 157
— —, —, Mal'cev's theorem on 75
locally cyclic 33
— dihedral 2-group 36
— finite 35—37, 92 ff.
— nilpotent groups with Max-*n* 166
— — — — Min-*n* 154
— normal and $\mathfrak{X}$-group 36
— soluble groups with Max-*n* 166
— — — — Min-*n* 154
— $\mathfrak{X}$-group 5

Mal'cev, theorem of 39, 41, 75
marginal subgroup 9, 111 ff.
maximal condition 37, 65
— —, groups with the — — locally 58, 72—73
— — on abelian subgroups 38, 85 ff., 131, 180
— — on finite subgroups 70
— — on normal subgroups 38, 158 ff.
— — on right ideals 163
— — on subnormal subgroups 38, 170—171, 180
— subgroup 17, 166—167
— normal subgroup 17

metabelian group, finitely generated 161
minimal condition 37, 68
— — on abelian subgroups 38, 85 ff., 131, 180
— — on normal subgroups 38, 146 ff
— — on p-subgroups 74
— — on subnormal subgroups 38, 170—171, 180 ff.
— —, π- 74, 135
— non-abelian 98
— non-Černikov 98
— non-nilpotent 98
— normal subgroup 17, 149 ff.
— subnormal subgroup 177 ff.
modular law, Dedekind's 39
monendomorphism 108
monolith, monolithic group 152
multiplicator, Schur's 103

n-generator group 30
$\tilde{N}$-group 167
nilpotent class 49
— group 49
— —, generalized 49
— — with Min 69
— — with Min-n 152
— length 168
— product 125
noetherian group 38
normal closure 42
— —, series of successive 173
— series 12
— subset 45
— system 10
normality, groups with transitive 174

operations on classes of groups 3
operator group 38

π-free 53
π-number 53
perfect group 7
— —, $\mathfrak{X}$- 26
periodic group 34
permutations, group of finitary 89
polycyclic 65
— -by-finite 65
poly-$\mathfrak{X}$ 4
presentation 31
—, finite 31
primary decomposition of group with Max-n 169—170
prime-thin direct product 135
product class 2
property of groups 1

quasicyclic group 67—68
— finite group 92

radicable group 69
radical class 20
— — generated by $\mathfrak{X}$ 20
— group 59
—, $\mathfrak{X}$- 18
rank, general 34
—, groups with finite 33
—, special 34
refinement 16
Reidemeister-Schreier Theorem 31
residual class 23
— — generated by $\mathfrak{X}$ 23
—, $\mathfrak{X}$- 18
residually $\mathfrak{X}$-group 4

SI-group 46
SI^*-group 46
SJ^*-group 46
SN-group 25
— — with Min 71
SN^*-group 46
Schmidt's Problem 92, 97
— —, dual of 123
Schreier Refinement Theorem 66, 184
Schur's Lemma 81
— pair 112 ff.
— theorem on central-by-finite groups 102
section 99
semi-regular permutation group 144
semi-simple class 23
— — group 26
— — —, $\mathfrak{X}$- 26
serial 12
series, 9—10
—, ascending 10—11
—, central 48
—, —, FC- 129
—, chief 17
—, —, strong 66
—, —, weak 66
—, complete 10
—, composition 16
—, —, of finite length 66, 171, 184
—, —, strong 66

series, composition, weak 66
—, descending 10–11
—, normal 12
— of successive normal closures 173
—, principal 17
—, subnormal 12
—, upper socle 151
—, χ- 13
simple, absolutely 16
—, characteristically 18
— groups, direct products of 178ff.
— —, embedding in 144ff.
— —, infinite 143ff.
—, strictly 17
— subgroup, subnormal 177ff.
socle, 149, 178
—, abelian 150
—, non-abelian 150
—, subnormal 180ff.
soluble group 45
— — with Max-n 160
— — with Min-n 152–153
soluble length 45
solvable group 45
subgroup theoretical class 9
— — property 9
subnormal 12
— index 173
— —, groups with bounded 173ff.
— intersection property 175
— join property 175
subsoluble group 22, 176
Šunkov, theorem of 93
supersoluble 66
—, locally 73, 156, 169

tensor product 54
thin 138
Three Subgroup Lemma 44
torsion group 34
transfer 101
transversal, right 30
triangulable 74
Trofimov number 127

unary operation 6
unimodular group 81
unitriangulable 74
universal class 7

variety 7
— generated by a class of groups 8
verbal mapping 8
— subgroup 8, 111ff.

Wielandt subgroup 177ff.
—, theorem of 59, 175
word, commutator 112
—, —, outer 112
—, concise 119
—, derived 119
—, lower central 113
—, verbose 119

$\mathfrak{X}$-group 1

Z-group 49
ZA-group 14, 49
ZD-group 49
Zassenhaus' Lemma 17
— theorem on soluble linear groups 78

721/2/72

Ergebnisse der Mathematik und ihrer Grenzgebiete

1. Bachmann: Transfinite Zahlen.
2. Miranda: Partial Differential Equations of Elliptic Type.
4. Samuel: Méthodes d'Algèbre Abstraite en Géométrie Algébrique.
5. Dieudonné: La Géométrie des Groupes Classiques.
7. Ostmann: Additive Zahlentheorie. 1. Teil: Allgemeine Untersuchungen.
8. Wittich: Neuere Untersuchungen über eindeutige analytische Funktionen.
11. Ostmann: Additive Zahlentheorie. 2. Teil: Spezielle Zahlenmengen.
13. Segre: Some Properties of Differentiable Varieties and Transformations.
15. Zeller/Beckmann: Theorie der Limitierungsverfahren.
16. Cesari: Asymptotic Behavior and Stability Problems in Ordinary Differential Equations.
17. Severi: Il teorema di Riemann-Roch per curve-superficie e varietà questioni collegate.
18. Jenkins: Univalent Functions and Conformal Mapping.
19. Boas/Buck: Polynomial Expansions of Analytic Functions.
20. Bruck: A Survey of Binary Systems.
23. Bergmann: Integral Operators in the Theory of Linear Partial Differential Equations.
25. Sikorski: Boolean Algebras.
26. Künzi: Quasikonforme Abbildungen.
27. Schatten: Norm Ideals of Completely Continuous Operators.
28. Noshiro: Cluster Sets.
29. Jacobs: Neuere Methoden und Ergebnisse der Ergodentheorie.
30. Beckenbach/Bellmann: Inequalities.
31. Wolfowitz: Coding Theorems of Information Theory.
32. Constantinescu/Cornea: Ideale Ränder Riemannscher Flächen.
33. Conner/Floyd: Differentiable Periodic Maps.
34. Mumford: Geometric Invariant Theory.
35. Gabriel/Zisman: Calculus of Fractions and Homotopy Theory.
36. Putnam: Commutation Properties of Hilbert Space Operators and Related Topics.
37. Neumann: Varieties of Groups.
38. Boas: Integrability Theorems for Trigonometric Transforms.
39. Sz.-Nagy: Spektraldarstellung linearer Transformationen des Hilbertschen Raumes.
40. Seligman: Modular Lie Algebras.
41. Deuring: Algebren.
42. Schütte: Vollständige Systeme modaler und intuitionistischer Logik.
43. Smullyan: First-Order Logic.
44. Dembowski: Finite Geometries.
45. Linnik: Ergodic Properties of Algebraic Fields.
46. Krull: Idealtheorie.
47. Nachbin: Topology on Spaces of Holomorphic Mappings.
48. A. Ionescu Tulcea/C. Ionescu Tulcea: Topics in the Theory of Lifting.
49. Hayes/Pauc: Derivation and Martingales.
50. Kahane: Séries de Fourier Absolument Convergents.
51. Behnke/Thullen: Theorie der Funktionen mehrerer komplexer Veränderlichen.
52. Wilf: Finite Sections of Some Classical Inequalities.

53. Ramis: Sous-ensembles analytiques d'une variété banachique complexe.
54. Busemann: Recent Synthetic Differential Geometry.
55. Walter: Differential and Integral Inequalities.
56. Monna: Analyse non-archimédienne.
57. Alfsen: Compact Convex Sets and Boundary Integrals.
58. Greco/Salmon: Topics in $\mathfrak{m}$-Adic Topologies.
59. López de Medrano: Involutions on Manifolds.
60. Sakai: C*-Algebras and W*-Algebras.
61. Zariski: Algebraic Surfaces.
62. Robinson: Finiteness Conditions and Generalized Soluble Groups, Part 1.
63. Robinson: Finiteness Conditions and Generalized Soluble Groups, Part 2.
64. Hakim: Topos annelés et schémas relatifs.
65. Browder: Surgery on Simply-Connected Manifolds.
66. Pietsch: Nuclear Locally Convex Spaces.
67. Dellacherie: Capacités et processus stochastiques.
68. Raghunathan: Discrete Subgroups of Lie Groups.